QUANTUM QUESTIONS

QUANTUM QUESTIONS

Mystical Writings of the World's Great Physicists

EDITED BY

Ken Wilber

WITH THE
RESEARCH ASSISTANCE OF

Ann Niehaus

SHAMBHALA
Boston
2001

Sʜᴀᴍʙʜᴀʟᴀ Pᴜʙʟɪᴄᴀᴛɪᴏɴꜱ, Iɴᴄ.
Horticultural Hall
300 Massachusetts Avenue
Boston, Massachusetts 02115
www.shambhala.com

9   8   7   6   5

Distributed in the United States by Random House, Inc., and in
Canada by Random House of Canada Ltd
Printed in the United States of America
⊗ This edition is printed on acid-free paper that meets the
American National Standards Institute z39.48 Standard.

Library of Congress Cataloging in Publication Data
Main entry under title:
Quantum Questions
1. Physics—Addresses, essays lectures.   2. Mysticism
—Addresses, essays, lectures.   3. Physics—Religious
aspects—Addresses, essays, lectures.   1. Wilber, Ken.
QC71Q36   1984   530   83-20332
ISBN 0-87773-266-3 (pbk.)
ISBN 0-394-72338-4 (pbk.: Random House)
ISBN 1-57062-768-1 (pbk.)

Front cover art: *Geflecht* © 1969 by Jurgen Peters

Dedicated to Terry

Contents

Planck

Pauli

Eddington

Preface to the 2001 Edition

I N THE MIDST of the growing interest in the relation between science and religion, it is always useful to return to the pioneering founders of modern physics and read what they themselves had to say on this most important topic: *Quantum Questions* is a compendium of virtually all of the significant writings from some of the greatest physicists the world has ever known.

The common tendency, when faced with the truly ultimate issues of existence, is to assume—or at least hope—that physics and mysticism would somehow converge on a similar set of answers, that physics would somehow support or even prove a mystical worldview. This, after all, has been the message of countless books, from *The Tao of Physics* to *The Dancing Wu-Li Masters*.

That simple conclusion, however, was not believed by any of the great physicists in this volume. From Einstein to Eddington, from Bohr to Planck, from Heisenberg to Pauli, they uniformly rejected that conclusion. They rejected the notion that physics proves or even supports mysticism, and *yet every one of them was an avowed mystic!*

How can that be? Very simply, they all realized that, at the very least, physics deals with the world of form, and mysticism deals with the formless. Both are important, but they cannot be equated. Physics can be learned by the study of facts and mathematics, but mysticism can only be learned by a profound change in consciousness. To confuse these two is to misunderstand and distort both science and spirituality.

As you will see in the following pages, all of these pioneering physicists believed that both science and religion, physics and spirituality, were necessary for a complete and full and integral approach to reality, but neither could be reduced to, or derived from, the other. (If you would like to follow up on these topics, I recommend starting with my book *A Theory of Everything: An Integral Vision for Business, Politics, Science, and Spirituality.*)

In these days when so many spiritual seekers feel that they need to rest their souls on the findings of physics, it is important that we listen to the true masters of physics as they point to the fundamental importance of both science and religion—without confusing their respective tasks and goals, yet finding them both as part of that All which only alone is.

Ken Wilber
Boulder, Colorado
Fall 2000

Preface

THE THEME OF THIS BOOK, if I may briefly summarize the argument of the physicists presented herein, is that modern physics offers no positive support (let alone proof) for a mystical worldview.
Nevertheless, every one of the physicists in this volume was a mystic.
They simply believed, to a man, that if modern physics no longer objects
to a religious worldview, it offers no positive support either; properly
speaking, it is indifferent to all that. The very compelling reasons why
these pioneering physicists did not believe that physics and mysticism
shared similar worldviews, and the very compelling reasons that they
nevertheless *all* became mystics—just that is the dual theme of this anthology. If they did not get their mysticism from a study of modern
physics, where *did* they get it? And why?

It is not my aim in this volume to reach the new age audience, who
seem to be firmly convinced that modern physics automatically supports
or proves mysticism. It does not. But this view is now so widespread, so
deeply entrenched, so taken for granted by New-Agers, that I don't see
that any one book could possibly reverse the tide. It was, I believe, with
every good intention that this "physics-supports-mysticism" idea was
proposed, and it was with very good intention that it was so rapidly and
widely accepted. But I believe these good intentions were misplaced, and
the results have been not just wrong but detrimental. If today's physics
supports mysticism, what happens when tomorrow's physics replaces it?
Does mysticism then fall also? We cannot have it both ways. As particle
physicist Jeremy Bernstein put it, "If I were an Eastern mystic the last
thing in the world I would want would be a reconciliation with modern
science, [because] to hitch a religious philosophy to a contemporary science is a sure route to its obsolescence." Genuine mysticism, *precisely* to
the extent that it is genuine, is perfectly capable of offering its own de-

fense, its own evidence, its own claims, and its own proofs. Indeed, that is exactly what the physicists in this volume proceed to do, without any need to compromise poor physics in the process.

No, the audience I would like to reach is the same audience these physicists wanted to reach: the orthodox, the established; the men and women who honestly believe that natural science can and will answer all questions worth asking. And so, in that orthodox spirit, I would simply ask, you of orthodox belief, you who pursue disinterested truth, you who—whether you know it or not—are molding the very face of the future with your scientific knowledge, you who—may I say so?—bow to physics as if it were a religion itself, to you I ask: what does it mean that the founders of your modern science, the theorists and researchers who pioneered the very concepts you now worship implicitly, the very scientists presented in this volume, what does it mean that they were, *every one of them, mystics?*

Does that not stir something in you, curiosity at least? Cannot the spirit of these pioneers reach out across the decades and touch in you that "still, small point" that moved them all to wonderment?

The last thing these theorists would want you to surrender is your critical intellect, your hard-earned skepticism. For it was exactly through a sustained use—not of emotion, not of intuition, not of faith—but a sustained use of the critical intellect that these greatest of physicists felt absolutely compelled to go beyond physics altogether. And as we will see in the following pages, they left a trail, clear enough, for all sensitive souls to follow.

K. W.
Muir Beach
Winter, 1983

QUANTUM QUESTIONS

1

Introduction:
Of Shadows and Symbols

BY KEN WILBER

BEYOND THE CAVE

PHYSICS AND MYSTICISM, physics and mysticism, physics and mysticism . . . In the past decade there have appeared literally dozens of books, by physicists, philosophers, psychologists, and theologians, purporting to describe or explain the extraordinary relationship between modern physics, the hardest of sciences, and mysticism, the tenderest of religions. Physics and mysticism are fast approaching a remarkably common worldview, some say. They are complementary approaches to the same reality, others report. No, they have nothing in common, the skeptics announce; their methods, goals, and results are diametrically opposed. Modern physics, in fact, has been used to both support and refute determinism, free-will, God, Spirit, immortality, causality, predestination, Buddhism, Hinduism, Christianity, and Taoism.

The fact is, every generation has tried to use physics to both prove and disprove Spirit—which ought to tell us something right there. Plato announced that the whole of physics was, to use his terms, nothing more than a "likely story," since it depended ultimately on nothing but the evidence of the fleeting and shadowy senses, whereas truth resided in the transcendental Forms beyond physics (hence "metaphysics"). Democritus, on the other hand, put his faith in "atoms and the void," since nothing else, he felt, had any existence—a notion so obnoxious to Plato that he expressed the strongest desire that all the works of Democritus be burned on the spot.

When Newtonian physics ruled the day, the materialists seized upon physics to prove that, since the universe was obviously a deterministic machine, there could be no room for free will, God, grace, divine intervention, or anything else that even vaguely resembled Spirit. This seemingly impenetrable argument, however, had no impact whatsoever on the spiritually-minded or idealistic philosophers. In fact, they pointed out, the second law of thermodynamics—which unequivocally announces that the universe is winding down—can mean only one thing: if the universe is winding down, something or somebody had to have previously wound it up. Newtonian physics doesn't disprove God; on the contrary, they maintained, it proves the absolute necessity of a Divine Creator!

When relativity theory entered the scene, the whole drama repeated itself. Cardinal O'Connell of Boston warned all good Catholics that relativity was "a befogged speculation producing universal doubt about God and his creation"; the theory was "a ghastly apparition of Atheism." Rabbi Goldstein, on the other hand, solemnly announced that Einstein had done nothing less than produce "a scientific formula for monotheism." Similarly, the works of James Jeans and Arthur Eddington were greeted by cheers from the pulpits all over England—modern physics supports Christianity in all essential respects! The problem was, Jeans and Eddington by no means agreed with this reception, nor in fact with each other, which prompted Bertrand Russell's famous witticism that "Sir Arthur Eddington deduces religion from the fact that atoms do not obey the laws of mathematics. Sir James Jeans deduces it from the fact that they do."

Today we hear of the supposed relation between modern physics and Eastern mysticism. Bootstrap theory, Bell's theorem, the implicate order, the holographic paradigm—all of this is supposed to prove (or is it disprove?) Eastern mysticism. In all essential respects it is simply the same story with different characters. The pros and cons strut their wares, but what remains true and unchanged is simply that the issue itself is extremely complex.

In the midst of this melange, then, it seemed a good idea to consult the founders of modern physics on what *they* thought about the nature of science and religion. What is the relation, if any, between modern physics and transcendental mysticism? Does physics bear at all on the issues of free-will, creation, Spirit, the soul? What *are* the respective roles of science and religion? Does physics even deal with Reality (capital R), or is it necessarily confined to studying the shadows in the cave?

This volume is a condensed collection of virtually every major statement made on those topics by the founders and grand theorists of modern (quantum and relativity) physics: Einstein, Schroedinger, Heisenberg, Bohr, Eddington, Pauli, de Broglie, Jeans, and Planck. While it would be asking too much to have all these theorists precisely agree with each other on the nature and relation of science and religion, nevertheless, I was quite surprised to find a very general commonality emerge in the worldviews of these philosopher-scientists. While there are exceptions (as we will see), certain strong and common conclusions were reached by virtually every one of these theorists. I will return to these general conclusions in a moment and state them more carefully and precisely, but by way of first approximation, we can say this: these theorists are virtually unanimous in declaring that modern physics offers no positive support whatsoever for mysticism or transcendentalism of any variety. (And yet they were *all* mystics of one sort or another! The reason for *that* will be one of the central questions of this section.)

According to their general consensus, modern physics neither proves nor disproves, neither supports nor refutes, a mystical-spiritual worldview. There *are* certain similarities between the worldview of the new physics and that of mysticism, they believe, but these similarities, where they are not purely accidental, are trivial when compared with the vast and profound differences between them. To attempt to bolster a spiritual worldview with data from physics—old or new—is simply to misunderstand entirely the nature and function of each. As Einstein himself put it, "The present fashion of applying the axioms of physical science to human life is not only entirely a mistake but has also something reprehensible in it."[1] When Archbishop Davidson asked Einstein what effect the theory of relativity had on religion, Einstein replied, "None. Relativity is a purely scientific theory, and has nothing to do with religion"—about which Eddington wittily commented, "In those days one had to become expert in dodging persons who were persuaded that the fourth dimension was the door to spiritualism."[2]

Eddington, of course, had (like Einstein) a deeply mystical outlook, but he was absolutely decisive on this point: "I do not suggest that the new physics 'proves religion' or indeed gives any positive grounds for religious faith. . . . *For my own part I am wholly opposed to any such attempt.*"[3] Schroedinger—who, in my judgment, was probably the greatest mystic in this group—was just as blunt: "Physics has nothing to do with it. Physics takes its start from everyday experience, which it continues by more subtle means. It remains akin to it, does not transcend it generically, it cannot enter into another realm."[4] The attempt to do so, he says, is simply "sinister": "The territory from which previous scientific attainment is invited to retire is with admirable dexterity claimed as a playground of some religious ideology that cannot really use it profitably, because its [religion's] true domain is far beyond anything in reach of scientific explanation."[5]

Planck's view, if I may summarize it, was that science and religion deal with two very different dimensions of existence, between which, he believed, there can properly be neither conflict nor accord, any more than we can say, for instance, that botany and music are in conflict or accord. The attempts to set them at odds, on the one hand, or "unify them," on the other, are "founded on a misunderstanding, or, more precisely, on a confusion of the images of religion with scientific statements. Needless to say, the result makes no sense at all."[6] As for Sir James Jeans, he was simply flabbergasted: "What of the things which are not seen which religion assures us are eternal? There has been much discussion of late of the claims of ["scientific support" for "transcendental events"]. Speaking as a scientist, I find the alleged proofs totally unconvincing; speaking as a human being, I find most of them ridiculous as well."[7]

Now it cannot be claimed that these men were simply unaware of the mystical writings of the East and West; that if they simply read *The Dancing Wu-Li Masters* they would all change their minds and pronounce physics and mysticism to be fraternal twins; that if they knew more about the details of the mystical literature they would indeed find numerous similarities between quantum mechanics and mysticism. On the contrary, their writings are positively loaded with references to the Vedas, the Upanishads, Taoism (Bohr made the yin-yang symbol part of his family crest), Buddhism, Pythagoras, Plato, Berkeley, Plotinus, Schopenhauer, Hegel, Kant, virtually the entire pantheon of perennial philosophers, and they still reached the above-mentioned conclusions.

They were perfectly aware, for instance, that a key tenet of the perennial philosophy is that in mystical consciousness subject and object become *one* in the act of knowing; they were also aware that certain philosophers claimed that Heisenberg's Uncertainty Principle and Bohr's Complementarity Principle supported this mystical idea, because, it was said, in order for the subject to know the object, it had to "interfere" with it, and that proved that the subject-object duality had been transcended by modern physics. *None of the physicists in this volume believed that assertion.* Bohr himself stated quite plainly that "the notion of complementarity does in no way involve a departure from our position as detached observers of nature. . . . The essentially new feature in the analysis of quantum phenomena is the introduction of a *fundamental distinction between the measuring apparatus and the objects under investigation* [his ital.]. . . . In our future encounters with reality we shall have to distinguish between the objective and the subjective side, to make a division between the two."[8,9] Louis de Broglie was even more succinct: "[It has been said that] quantum physics reduces or blurs the dividing region between the subjective and the objective, but there is . . . some misuse of language here. For in reality the means of observation clearly belong to the objective side; and the fact that their reactions on the parts of the external world which we desire to study cannot be disregarded in microphysics neither abolishes, nor even diminishes, the traditional distinction between subject and object."[10] Schroedinger—and keep in mind that these men firmly acknowledged that in mystical union subject and object are one, they simply found no support for this idea whatsoever in modern physics—stated that "the 'pulling down of the frontier between observer and observed' which many consider [a] momentous revolution of thought, to my mind seems a much overrated provisional aspect without profound significance."[11]

Accordingly, for the reasons that these theorists rejected the "physics-supports-mysticism" view, we will have to look elsewhere than the alleged fact that they were unacquainted with mystical literature or experience. And even if their knowledge of, say, Taoism, could be shown to be deficient, their critique would still, I believe, be absolutely valid. Further, this critique (which I will present in a moment) is not affected one way or another by any particular advances in physics; it is a logical critique that cuts at right angles to any possible new discoveries. This critique is simple, straightforward, and profound; at one stroke, it cuts across virtually everything written on the supposed parallels between physics and mysticism.

Briefly, the critique is this. The central mystical experience may be fairly (if somewhat poetically) described as follows: in the mystical consciousness, Reality is apprehended directly and immediately, meaning without any mediation, any symbolic elaboration, any conceptualization, or any abstractions; subject and object become one in a timeless and spaceless act that is beyond any and all forms of mediation. Mystics universally speak of contacting reality in its "suchness," its "isness," its "thatness," without any intermediaries; beyond words, symbols, names, thoughts, images.

Now, when the physicist "looks at" quantum reality or at relativistic reality, he is *not* looking at the "things in themselves," at noumenon, at direct and nonmediated reality. Rather, the physicist is looking at *nothing but a set of highly abstract differential equations*—not at "reality" itself, but at mathematical symbols of reality. As Bohr put it, "It must be recognized that we are here dealing with a *purely symbolic procedure.* . . . Hence our whole space-time view of physical phenomena depends ultimately upon these abstractions."[12] Sir James Jeans was specific: in the study of modern physics, he says, "we can never understand what events are, but must limit ourselves to describing the patterns of events in mathematical terms; no other aim is possible. Physicists who are trying to understand nature may work in many different fields and by many different methods; one may dig, one may sow, one may reap. But the final harvest will always be a sheaf of mathematical formulae. These will *never* describe nature itself. . . . [Thus] our studies can never put us into contact with reality."[13]

What an absolute, radical, irredeemable difference from mysticism! And this critique applies to any type of physics—old, new, ancient, modern, relativistic, or quantum. The very nature, aim, and results of the approaches are profoundly different: the one dealing with abstract and mediate symbols and forms of reality, the other dealing with a direct and nonmediated approach to reality itself. To even claim that there are direct and central similarities between the findings of physics and mysticism is necessarily to claim the latter is fundamentally a merely symbolic abstraction, because it is absolutely true that the former is exactly that. At the very least, it represents a profound confusion of absolute and relative truth, of finite and infinite, of temporal and eternal—and that is what so repelled the physicists in this volume. Eddington, as usual, put it most trenchantly: "We should suspect an intention to reduce God to a system of differential equations. That fiasco at any rate [must be] avoided. However much the ramifications of [physics] may be extended by further scientific discovery, they cannot from their very nature trench on the background in which they have their being. . . . We have learnt that the exploration of the external world by the methods of physical science leads not to a concrete reality but to a *shadow world of symbols,* beneath which those methods are unadapted for penetrating."[14]

Physics, in short, deals with—and can only deal with—the world of shadow-symbols, not the light of reality beyond the shadowy cave. Such, as a brief first approximation, is the general conclusion of these theorists.

But why, then, did *all* of these great physicists embrace mysticism of one sort or another? Obviously, there is *some* type of profound connection here. We have seen that this connection does *not* lie, according to these theorists, in a similarity of worldviews between physics and mysticism, nor a similarity in aim or results; between shadow and light there can be no fundamental similarity. So what forced so many physicists out of the cave? What, in particular, did the *new* physics (quantum and relativistic) tell these physicists that the old physics failed to mention? What, in brief, was the crucial difference between the old and new physics, such that the latter tended much more often to be conducive to mysticism?

There is, once again, a general and common conclusion reached by the majority of the theorists in this volume, and best elucidated by Schroedinger and Eddington. Eddington begins with the acknowledged fact that physics is dealing with shadows, not reality. Now the great difference, he says, between the old and the new physics is not that the latter is relativistic, nondeterministic, four-dimensional, or any of those sorts of things. The great difference between old and new physics is both much simpler and much more profound: both the old and the new physics were dealing with shadow-symbols, *but the new physics was forced to be aware of that fact*—forced to be aware that it was dealing with shadows and illusions, not reality. Thus, in perhaps the most famous and oft-quoted passage of any of these theorists, Eddington eloquently states: "In the world of physics we watch a shadowgraph performance of familiar life. The shadow of my elbow rests on the shadow table as the shadow ink flows over the shadow paper. . . . The frank realization that physical science is concerned with a world of shadows is one of the most significant of recent advances."[15] Schroedinger drives the point home: "Please note that the very recent advance [of quantum and relativistic physics] does not lie in the world of physics itself having acquired this shadowy character; it had ever since Democritus of Abdera and even before, *but we were not aware of it; we thought we were dealing with the world itself.*"[16] And Sir James Jeans summarizes it perfectly, right down to the metaphor: "The essential fact is simply that *all* the pictures which science now draws of nature, and which alone seem capable of according with observational fact, are *mathematical* pictures. . . . They are nothing more than pictures—fictions if you like, if by fiction you mean that science is not yet in contact with ultimate reality. Many would hold that, from the broad philosophical standpoint, the outstanding achievement of twentieth-century physics is not the theory of relativity with its welding together of space and time, or the theory of quanta with its present apparent negation of the laws of causation, or the dissection of the atom with the resultant discovery that things are not what they seem; it is the general recognition that we are not yet in contact with ultimate reality. We are still imprisoned in our cave, with our backs to the light, and can only watch the shadows on the wall."[17]

There is the great difference between the old and new physics—both are dealing with shadows, but the old physics didn't recognize that fact. If you are *in* the cave of shadows and don't even know it, then of course you have no reason or desire to try to escape to the light beyond. The shadows appear to be the whole world, and no other reality is acknowledged or even suspected—this tended to be the philosophic effect of the old physics. But with the new physics, the shadowy character of the whole enterprise became much more obvious, and sensitive physicists by the droves—including all of those in this volume—began to look beyond the cave (and beyond physics) altogether.

"The symbolic nature of physics," Eddington explains, "is generally recognized, and the scheme of physics is now formulated in such a way as to make it almost self-evident that it is a partial aspect of something wider." However, according to these physicists, about this "something wider" physics tells us—and can tell us—nothing whatsoever. It was exactly this radical failure of physics, and not its supposed similarities to mysticism, that paradoxically led so many physicists to a mystical view of the world. As Eddington carefully explains: "Briefly the position is this. We have learnt that the exploration of the external world by the methods of physical science leads not to a concrete reality but to a shadow world of symbols, beneath which those methods are unadapted for penetrating. Feeling that there must be more behind, we return to our starting point in *human consciousness*—the one centre where more might become known. There [in immediate inward consciousness] we find other stirrings, other revelations than those conditioned by the world of symbols. . . . Physics most strongly insists that its methods do not penetrate behind the symbolism. Surely then that mental and spiritual nature of ourselves, known in our minds by an intimate contact transcending the methods of physics, supplies just that . . . which science is admittedly unable to give."[18]

To put it in a nutshell: according to this view, physics deals with shadows; to go beyond shadows is to go beyond physics; to go beyond physics is to head toward the meta-physical or mystical—and *that* is why so many of our pioneering physicists were mystics. The new physics contributed nothing positive to this mystical venture, except a spectacular failure, from whose smoking ruins the spirit of mysticism gently arose.

A CLOSER LOOK

I should like, in this section, to look more closely at the relation between science and religion, their nature, methods, and domains. It must be emphasized, however, that in this section, unlike the previous section, I am not necessarily representing the views of the physicists in this volume; these are more or less my own ideas, which will, I believe, help clarify the issues in this anthology. And while many, perhaps most, of the physicists included herein would probably agree with most of what I have to say, nevertheless we are now dealing, not with generalities or commonalities, but with specific details and terminology, about which each physicist had his own particular and often idiosyncratic views. I will often indicate the points with which the various physicists would agree, and those points with which they would probably disagree.

There is, first of all, the very meaning of the word "science." We are, of course, free to define "science" any way we wish, as long as we are consistent, and, in fact, much of this "science-and-religion" argument consists of nothing more than a jockeying for definitions selected in advance to produce precisely the conclusion desired. Thus, for instance, if you define science simply as "knowledge," then contemplative religion becomes a form of science—becomes, in fact, the highest science (this approach is often taken by present-day Eastern masters, who continually speak of the science of yoga, the science of meditation, the science of creative intelligence, and so on). Physics then becomes a branch of that all-encompassing Tree of Science, and we're off and running with *The Medium, the Mystic, and the Physicist.*

On the other hand, if you define science as "empiricalsensory knowledge, instrumentally validated," then virtually all forms of religion become non scientific. You then have one of two paths open: 1) view religion as a perfectly valid form of *personal* faith, values, and belief not open to scientific scrutiny—these are said to be two different but equally legitimate domains between which there can properly be neither conflict nor compromise nor parallels (this view was pioneered by Kant and Lotze, and has many adherents to this day, including some of the physicists in this volume, such as Planck, Einstein, and Eddington); 2) view religion as nonscientific in the purely pejorative sense, as a superstitious relic of magical and primitive thinking (Comte), or a defense mechanism expiating guilt and anxiety (Freud), or an opaque ideology institutionalizing alienation (Marx), or a debilitating projection of men's and women's inward and humanistic yearnings (Feuerbach), or a purely private emotional affair, harmless in itself but not deserving the title "knowledge" (Quine, Ayer, and the positivists).

Now *all* of that confusion, you see, rests in large measure on how you define "science." The issues are so complex and subtle that if we don't specify precisely what we mean by "science" (and later, by "religion"), then statements about the relation between the two become silly at best, sinister at worst. Personally, I am now at the point that, if a popular writer makes some sweeping statement about the "new science" and "spirituality," I have no idea whatsoever about what they might mean, and all I feel certain of is that they don't, either.

Since this entire anthology is, in fact, devoted to the themes of "science and religion," I don't see that we have any choice except to examine very carefully what we mean, or *can* mean, by the word "science" and the word "religion." My somewhat dreary editorial task, then, for the next few pages, is to play the role of linguistic analyst, that most banal of all philosophic activities. I shall try to make the operation as painless as possible.

Start with "science." As I said, we are free to define "science" any way we wish, as long as we are consistent. But it seems to me that at the very least we must distinguish between the *method* of science and the *domain* of science. The *method* of science refers to the ways or means that whatever it is we call science manages to gather facts, data, or information, and manages to confirm or refute propositions vis à vis that data. Method, in other words, refers to ways in which "science" (still unspecified) manages to gather knowledge.

Domain, on the other hand, simply refers to the types of events or phenomena that become, or can become, objects of investigation by whatever it is we mean by science. "Method" refers to the epistemology of science, while "domain" refers to its ontology.

Let me give a crude analogy. Say we are exploring Carlsbad Caverns in the dark of night. We take a flashlight with us—that is our means or our method of gaining knowledge (or of "shedding light" on the various caves), and the caves are the different objects or domains that we will investigate and illuminate with our methodology, with our flashlight. One cave might contain buried treasure of gems and gold, another might contain nothing but mud and bats—the point is that the same flashlight might discover very different types of objects, and we don't want to confuse these objects simply because the same flashlight was used to find them.

Instead of asking vaguely "What is science?", let us therefore ask "What is a scientific *method?*" and "What is a scientific *domain?*" As for scientific method, general science texts seem to be in agreement: a method of gaining knowledge whereby *hypotheses* are *tested* (instrumentally or experimentally) by reference to *experience* ("data") that is potentially *public,* or open to *repetition* (confirmation or refutation) by peers. In bare essentials, it means that the scientific method involves those knowledge-claims open to *experiential* validation or refutation.

Notice that this definition—which we will accept for the moment—correctly makes no reference to the domain or objects of the scientific method. If there is a way to test a knowledge-claim *in whatever domain* by appeal to open experience, then that knowledge can properly be called "scientific."

This definition, correctly I believe, does *not* say that only sensory or physical objects are open to scientific investigation—that would be like claiming that our flashlight can be used in only one cave. There is nothing in that definition that prevents us from legitimately applying the term "scientific" to certain specifiable knowledge-claims in the realms or domains of biology, psychology, history, anthropology, sociology, and spirituality. Indeed, that is exactly what the Germans mean by "geist-science," the science of mental and spiritual phenomena, and what we Americans mean by the human or social sciences.

The point is that because this definition correctly concerns only method and makes no reference to object-domains, the dividing line between "scientific" and "non scientific" is *not* between physical and metaphysical; the dividing line is between experientially testable and nontestable (or merely dogmatic) pronouncements, the former being exposed to confirmation/refutation based on open experience, the latter being based on evidence no more substantial than the "because-I-tell-you-so" variety. If "science" were restricted to "physical-sensory" object-domains, then mathematics, logic, psychology, and sociology could not be called "scientific," in that the central aspects of those domains are non-sensory, non-empirical, non-physical, or meta-physical occasions.

There is, for instance, a way to *test* the truth-value of a mathematical theorem, but this test is based, not on *sensory* evidence, but on *mental* evidence, namely, the inward *experience* of the mental *coherence* of the train of logical propositions, an inwardly-experiential coherence that can be checked by the minds of other equally trained mathematicians, an inwardly-experiential coherence (not correspondence) that has nothing to do with physical-sensory evidence. (The correspondence, or lack of it, can also be tested by reference to evidence, either mental or sensory as the case requires.) The point is that "test by experiential evidence" does *not* mean merely "test by physical-sensory evidence" (a point we will soon elaborate), and that is exactly why mathematics, logic, psychology, and so forth are properly called "sciences."[19]

Having seen that "scientific method" applies to experientially testable knowledge-claims as opposed to nontestable, dogmatic proclamations (which *may* be valid, but on grounds we will have to call by something other than the term "scientific"), we can now ask, "To what *domain(s)*, then, is the scientific method applicable?" But let us ask first, "What domains *are* there?" That is, what realms of experience, or modes of being, or aspects of reality are even available, in the first place, to which the "scientific method" may or may not be applicable? In other words, what *ontology* shall we accept? How many caves are there in the universe that we may go exploring with our flashlight?

I am not going to make a long drawn-out argument over this; for the purposes of this presentation, I shall simply assume the basic ontology of the perennial philosophy; specifically, as summarized by Lovejoy, Huston Smith, René Guénon, Marco Pallis, Frithjof Schuon, et al., and as embraced (in whole or in part) by modern thinkers such as Nicolai Hartmann, Samuel Alexander, Whitehead, Aurobindo, Maritain, Urban, etc. Nor am I going to haggle over terms; God, Godhead, Absolute, Ultimate, Being, Spirit, life, consciousness, psyche, soul—those terms can mean pretty much whatever you want them to. My purpose lies in a different direction.

Here, then, is our working ontology—the so-called "Great Chain of Being" (see diagram).

Running up the diagram, I have appended a general name to each domain; running down the diagram, I have listed a representative discipline that generally (but not necessarily exclusively) takes as its object of study that particular domain. The numbers simply refer to the levels, and the letters I will explain in a moment. I might also mention that some versions of the Great Chain give anywhere from three to twenty or more levels; this simple five-level scheme will adequately serve our purposes.

The general meaning of the terms "matter," "life," and "mind" might be fairly obvious, but let me say a word or two about "soul" and "spirit." The soul-realm, as I will use the term, refers to the realm of Platonic Forms, archetypes, personal deity-forms (yidam, ishtadeva, archangelic patterns, and so forth). In the soul-realm, there is still some sort of subtle subject-object duality; the soul apprehends Being, or communes with God, but there still remains an irreducible boundary between them. In the realm of spirit (level 5), however, the soul *becomes* Being in a nondual state of radical intuition and supreme identity variously known as gnosis, nirvikalpa samadhi, satori, kensho, jnana, etc. In the soul-realm, the soul and God commune; in the spirit-realm, both soul and God unite in Godhead, or absolute spirit, itself without exclusive boundaries anywhere.

Already, however, we run into grave semantic difficulties with the word "spirit," for there is virtually no way we can discuss the realm of spirit without involving paradox. Spirit itself is not paradoxical; it is, strictly speaking, beyond all characterization and qualification whatsoever (including that one). Because spirit is, so to speak, the ultimate limit of the nested hierarchy of Being, it enters our verbal formulations in apparently contradictory or paradoxical ways (as Kant, Stace, Nagarjuna, and others have pointed out). This becomes problematic, however, *only* if we forget to include *both* sides of the paradox in our verbal formulations.

Let me give a few examples. Notice that each level in the Great Chain *transcends but includes* its predecessor(s). That is, each higher level contains functions, capacities, or structures not found on, or explainable solely in terms of, a lower level. The higher level does not violate the principles of the lower, it simply is not exclusively bound to or explainable by them. The higher transcends but includes the lower, and *not vice versa,* just as a three-dimensional sphere includes or contains two-dimensional circles, but not vice versa. And it is this "not-vice-versa" that establishes and constitutes nested hierarchy. Thus, for example, life transcends but includes matter, and not vice versa: biological organisms contain material components, but material objects do not contain biological components (rocks don't genetically reproduce, etc.). This is also

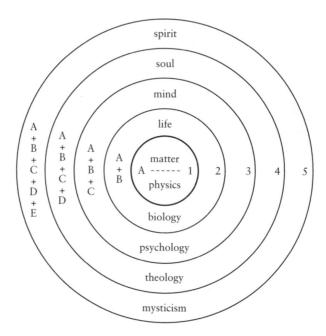

why, for example, in the study of biology one uses physics, but in the study of physics one does not use biology.*

Thus, life transcends but includes matter; mind transcends but includes life; soul transcends but includes mind; and spirit transcends but includes soul. At that point, however, asymptotic at infinity, we have reached a paradoxical limit: spirit is that which transcends everything *and* includes everything. Or, in traditional terms, spirit is both completely transcendent to the world and completely immanent in the world—and there is the most notorious (and unavoidable) paradox of spirit.

On the one hand, then, spirit is the *highest* of all possible domains; it

---

*How does the Great Chain metaphor relate to the shadow and cave metaphor? Very simply: the *levels* of the Great Chain are the *levels of the shadow-objects* in the cave, for some shadow-objects are obviously *closer* to the opening than others, which constitutes the *hierarchy* of the levels. The levels of the Great Chain are thus levels of decreasing shadow and increasing light, culminating at the opening to spirit (Level 5), whereupon we realize there was always and *only* spirit, even at the lowest levels (although this can only be *realized* at the highest.) This does not mean that the levels of manifestation are pure illusion or pure unreality, for they are all manifestations of Being and, therefore, bathe, in various degrees, in its glory. It is just that the higher levels, being closer to the opening, need (as Bradley put it) less and less supplementation to pass into the Absolute. The fact that *all* things are God, but is the Summit of all realms, the Being beyond all beings. It is the domain that is a subset of no other domain, and thus preserves its radically transcendental nature. On the other hand, since spirit is all-pervading and all-inclusive, since it is the set of all possible sets, the Condition of all conditions and the Nature of all natures, it is not properly thought of as a realm set apart from other realms, but as the Ground or Being of *all* realms, the pure *That* of which all manifestation is but a play or modification. And thus spirit preserves (paradoxically) its radically immanent nature.

Now I labor on this apparently trivial point for what is really a very important reason. Because spirit can legitimately be referred to as both perfectly transcendent and perfectly immanent, then if we aren't extremely careful which meaning we wish to convey, we can play fast and loose with statements about what is or is not the realm of spirit. Thus, for example, if we emphasize solely the transcendental nature of spirit, then religion (and spirit) are obviously "out of this world" and have absolutely nothing in common with earth-bound science. Any attempt to identify spirit with the manifest world of nature is, in this truncated view, charged with the ugly epithet of "featureless pantheism," and theologians are all in a tither to explain that "dragging God into the finite realm" supposedly abolishes all values and actually destroys any meaning we could attach to the word "God" or "spirit."

On the other hand, if we commit the equal but opposite error and emphasize solely the immanent nature of spirit, then not only are science and religion compatible, but science becomes a subset of religion, and "The more we know of things [science], the more we know of God [religion]" (Spinoza). Attempts to place God or Spirit in any sort of transcendental "realm beyond" are met with howling charges of "dogmatism" and "nonsensicality," and all congratulate themselves on solving the transcendental Mystery, whereas all they have done is ignore it.

Much of this confusion would evaporate if we (1) acknowledge the necessary paradoxicality of verbal formulations of spirit, and (2) simply indicate which aspect of spirit—transcendent or immanent—we mean at any given time. This is not a philosophical nicety; it is an absolutely crucial prerequisite to making any meaningful statement about the role and relation of science and religion.

For my part, then, when I wish to refer to spirit in its transcendental

---

some things are more God than others—that is another version of the paradoxicality of Being.

aspect—as the highest dimension or summit of being—I will use "spirit" or "spiritual realm" with a small "s," to indicate that the spiritual realm (level 5) is a realm that in some very significant ways is *different* from, or *transcendent* to, the realms of matter, life, mind, and soul. Specifically, I mean this: we said that each level transcends but includes its predecessor(s). If matter (level 1) has the characteristics of A, then biological life (level 2) can be represented as A + B, where B stands for all those capacities found in living organisms but not in inanimate matter (such as food consumption, metabolism, sex, motor functions, and so on). Mind (level 3) is then A + B + C, where C represents all those capacities found in psychological systems but not in biological or material systems (such as ideas, concepts, values, insights, and so on). Likewise, the soul-realm is A + B + C + D; and the spiritual realm (with a small "s") is A + B + C + D + E. Thus, when we speak of "exploring the spiritual realm" or "the characteristics of the spiritual realm," we mean exactly those functions, capacities, and aspects (represented by E) that are found in the spiritual realm and nowhere else (such as *jnana, nirvikalpa samadhi, nirguna Brahman,* and so on).

If we were allowed to speak of the "science of spirit" (we haven't yet addressed that issue), all we would mean is the "scientific investigation of those events that constitute class E." In that sense, spiritual science would definitely be significantly different from, but not at all antagonistic towards, physical science (study of class A), biological science (study of class B), psychological science (study of class C), and so on. Nothing whatsoever would be gained by trying to mix or confuse these sciences, or claim they are all "really one," or lump them together indiscriminately—that, again, would be like claiming that the gold in one cavern is the same as the mud in another because the same flashlight discovered both.

Now, when I wish to refer to the all-pervading, all-embracing, radically *immanent* aspect of spirit, I will use "Spirit" with a capital "S," to indicate that Spirit is not the highest level among other levels but rather is the Ground or Reality of *all* levels, and thus could have no specific qualities or attributes itself, other than being the "isness" (tzu jan) or "suchness" or "thatness" (tathata) of all possible and actual realms—in other words, the unqualifiable Being of all beings, not the qualifiable being of any particular beings, and certainly not class E as opposed to class A, B, C, or D. (In the diagram of the Great Nest of Being, Spirit is represented, not by level 5, but by the paper on which the entire diagram is drawn).*

As regards Spirit (not spirit), the important point is that Spirit is neither One nor Many, neither infinite nor finite, neither whole nor part—for *all* of those are supposed *qualifications* of Spirit, and thus could at best apply to spirit, not Spirit. This is exactly the Buddhist doctrine of *sunyata* ("nonqualifiability"), the negation of all negations. And in particular notice that Spirit is *not* One, not Wholeness, not Unity—*neti, neti*—for all of those are dualistic concepts, possessing meaning only in contrast to their opposites.

Now there is a legitimate meaning to, for example, "Wholeness"—namely, the sum Totality of everything in existence, levels 1–5. But that Wholeness or Totality, it must be emphasized, has precisely nothing to do with Spirit, which is radically, completely, absolutely, and equally immanent in and as every single particular anywhere in existence. Thus, seven things do not contain more Spirit than three things, and wholeness is not more Real than partialness. "Wholeness" does have an important applicability on the transcendental side of the paradox—for example, any biological object possesses more wholeness than any material object, and thus *is closer to the spiritual realm but not closer to Spirit.* I mention this so that we don't fall into the positivistic error of equating

---

*This paradox is exactly why most of the physicists in this volume will talk about some sort of "unity" between physics and mysticism (or the realms of matter and spirit), and yet often in the same sentence completely deny it. What they are doing, consciously or unconsciously, is reflecting the paradoxical nature of spirit/Spirit. As spirit (small s), it is the *highest* dimension (and therefore quite divorced from physics), and as Spirit (large S), it is the *common* Ground (and therefore "underlying" physics in a "unitary" fashion). Thus Eddington, as summarized by Cohen: "Professor Eddington's main thesis, so far as it has bearing upon religion, is the existence of two worlds, one to which scientific 'laws' apply, and another world, to which scientific laws have no application. But there is, he admits, 'a kind of unity between the material and the spiritual worlds . . . but to those who have any intimate acquaintance with laws of chemistry and physics [and here comes the paradoxical denial], the suggestion that the spiritual world could be ruled by laws of an allied [or so-called parallel] character, is as preposterous as the suggestion that a nation could be ruled by the laws of grammar.' " This type of paradoxicality is rampant in the works of Einstein, Eddington, Schroedinger, Bohr, Heisenberg—indeed, in virtually all of the theorists in this volume. That, as I believe, is exactly as it should be; problems arise only if we ignore or forget that inherent paradoxicality. I am simply trying to make it conscious and explicit, so it doesn't befuddle an already difficult enough situation.

Spirit with Totality or Wholeness, an error that, it seems to me, is quite popular nowadays, and under whose auspices many an outrageous philosophical sleight-of-hand has been perpetrated.*

I think we have enough tools now to return to our original questions and attempt closure.

What do we mean by "religion"?

In *A Sociable God* I presented eight equally legitimate uses of the term "religion"; for our much simpler purposes, we can say that the type of religion we are discussing in this volume is that which has—or claims to have—direct access to levels 4 and 5 (and especially 5). The question then becomes, does that type of religion (or spirituality) deserve the status of knowledge? Can it claim *valid* knowledge? Or even more specifically, does it deserve the status of *scientific* knowledge? From our previous discussion, we know that what the question really means is this: are religious phenomena (phenomena of levels 4 and 5) such that they can become a proper *domain* for the scientific *method*?

My own conclusion is that all domains (levels 1–5) contain certain features or deep structures that are open to scientific investigation, because *all* domains are open to *experiential* disclosure. There is religious *experience* just as certainly as there is psychological experience and sensory experience. In that sense, we can speak of the science of religion just as legitimately as we speak of the science of psychology, biology, or physics.

Now by "religious experience" I mean the direct apprehension, in consciousness, of those phenomena we have called class D and class E, or the domains of soul and spirit. The central features of those domains are not only experienceable, they are *public,* because consciousness can be *trained* to apprehend those domains (this training is called meditation or contemplation), and a trained consciousness is a public, shareable, or intersubjective consciousness, or it couldn't be trained in the first place. Simply because religious experience is apprehended in an "interior" fashion does not mean it is merely private knowledge, any more than the fact that mathematics and logic are seen inwardly, by the mind's eye,

---

*This is why Zen, for example, emphatically denies that Spirit is one, or whole, or an underlying unity or identity. As D. T. Suzuki put it: "Followers of identity are to be given the warning: they are ridden by concepts" *(Zen and Japanese Culture).* Zen says, if anything, that Spirit is "not-two, not-one!"

makes them merely private fantasies without public import. Mathematical knowledge is public knowledge to all equally trained mathematicians; just so, contemplative knowledge is public knowledge to all equally trained contemplatives. The preposterous claim that all religious experience is private and noncommunicable is stopped dead by, to give only one example, the *transmission* of Buddha's enlightenment all the way down to the present-day Buddhist masters.

This does not mean that all so-called "religious knowledge" passes the scientific (experiential and public) test. Dogmatic assertions, idiosyncratic preferences, personal and intentionally private beliefs, and nontestable theological claims—these may or may not be valid, but they are not scientifically demonstrable or refutable; they are, that is, nonscientific or nontestable knowledge-claims. On the other hand, virtually all the Eastern texts on meditation and yoga, and virtually all the Western texts on contemplation and interior prayer, can legitimately be called *scientific treatises* dealing (principally) with levels 4 and 5; they contain rules and *experiments,* which, if followed correctly, disclose to consciousness phenomena (or data) of the classes we have called D and E, phenomena that can be as easily checked with (confirmed or refuted by) equally trained peers as geometric theorems can be checked with (confirmed or refuted by) other equally trained mathematicians.

What about the conflict or battle between science and religion?

There *is* a real, genuine, and important battle here, I believe, but it is not properly stated as a battle between science and religion.

To begin with, we have seen that there is a difference between "domain" and "method," and thus we are really dealing with two completely different scales, so to speak. On the one hand, there are the natural and important differences between the lower and upper domains of existence. On the other hand, there are the natural and important differences between genuine or verifiable and dogmatic or nonverifiable knowledge-claims.

Unfortunately, when these scales are confused or equated, then science comes to mean "lower *and* genuine," and religion comes to mean "upper *and* nonsensical." The battle, thus stated, can *never* be resolved, because both parties are half-right and half-wrong. Properly speaking, there is no battle whatsoever between the lower and upper dimensions of reality (since the latter transcend but include the former). There is, however, a very real battle between genuine versus nonsensical knowledge-claims, but this battle is *not* a battle between lower and upper domains of existence. It is a battle that reappears on *every* realm of existence (levels 1–5) and concerns knowledge-claims that are open to experiential test versus those that are dogmatic and nonverifiable (or nonrefutable).

Thus, if by "science" you mean the study of the lower, base, or natural levels of existence (usually 1/2/3), and if by "religion" you mean an approach to the upper, higher, or "supernatural" levels (usually 4/5), then the only *real* battle is between genuine science and bogus science, and between genuine religion and bogus religion ("genuine" meaning "experientially verifiable/refutable"; "bogus" meaning "dogmatic, nonexperiential, nonverifiable/refutable"). *There is bogus or pseudo-science just as much as there is bogus or pseudo-religion,* and the only worthwhile battle is between genuine and bogus, not between science and religion.

Accordingly, both genuine science and genuine religion are allied against pseudo, nonexperientially grounded, dogmatic knowledge-claims (which infect all domains), which is why, at this point, we can just as easily refer to this methodological alliance as the science of physics, the science of biology, the science of psychology, and the science of religion (or spirituality). Here "science" refers not to any particular domain, high or low, but to a methodology based on experiential evidence and not dogmatic assertions, a methodology we want to apply to all genuine knowledge-claims on all levels; this is what we mean by the terms "higher" or "spiritual" or "geist" *sciences.* In no case, however, is there a genuine battle between science and religion, only a battle between experiential science and religion versus dogmatic science and religion.

Are the methods of the mental or spiritual sciences the same as those of the physical sciences?

Yes and no. Yes, in that the central methodological criterion—namely, that all knowledge-claims ultimately be settled on the basis of direct appeal to experience—is identical in all the genuine sciences, physical, biological, psychological, and spiritual. No, in that each domain has quite different characteristics, and thus the actual application of the scientific method in each domain takes on the form, as it were, of that domain.

For example, one of the dominant characteristics of the physical realm is its extension in space-time. The easiest way to deal with extension is to *measure* it; thus, measurement is very prominent in the physical sciences (this aspect of the physical sciences was discovered independently by Kepler and Galileo, in 1605, and so they are properly referred to as the fathers of modern physical science). By the time we get to the mental-psychological level, however, quantity and extension largely give way to quality and intention; therefore, quantitative measurement, although still applicable in certain areas, is not nearly so prominent. A typical knowledge-claim in the physical sciences is, "A proton has two thousand times the mass of an electron," whereupon we proceed to test the claim through complicated instrumental procedures. On the other hand, a typical knowledge-claim in the mental realm is, "The meaning of *Hamlet* is such and such," which we then *test* in the hermeneutic circle (or the intersubjective realm of communicative exchange) of those who have read and studied *Hamlet*. Bad interpretations can be rebuffed by the hermeneutic circle, thus assuring a quasi-objective status for all genuine truth-claims. But here we are not so much judging extension as we are intention, so measurement plays a minor role.

Likewise, a typical knowledge-claim in the spiritual realm is, "Does a dog have Buddha-nature?" There is a specific, repeatable, verifiable, experiential test and answer to that question—a bad answer can most definitely be refuted—but it has virtually nothing to do with physical measurement or mental intentionality.[20]

This overall approach, then, assures us of a unity-in-diversity of the knowledge quest: a unity in methodological criteria, or a unity in knowledge itself, underlying a diversity in its objects, or a diversity in its particular applications. Put somewhat poetically: unity in *knowledge* underlying diversity of *phenomena*. I say "poetically," because if we push that statement very hard, it will collapse in paradox (simply because it ultimately ascribes to Spirit the qualification of "underlying unity," which violates *sunyata*). But let us temporarily ignore that in order to ask the central question of this anthology: with reference to the actual data or phenomena of physics and mysticism, are there any important parallels? In other words,

Are there any significant parallels between the phenomena disclosed by physics and those disclosed by mysticism?

Here we are not discussing the abstract, central criteria of all genuine sciences, whether physical or psychological or mystical—we have already said that those share a central form. We are discussing the findings, the results, the data, the phenomena of the physical and mystical sciences, and asking whether *those* share any significant parallels. And there, whether we define mysticism as knowledge of spirit or as knowledge of Spirit, the answer is still "None (or at best, a few rather trivial ones)." *This is exactly the same conclusion we reached in the first section of this essay,* the conclusion that reflected the common or majority agreement of the physicists included in this volume, although we arrived at it by entirely different means. Then we took a rather steep or drastic approach, following the course of most of the physicists themselves. In light of our more extended discussion in this section, we can reach the same conclusion by a slower yet steadier route.

First, if by mysticism we mean a direct and experiential knowledge of the spiritual realm (level 5), then of course there will be some sort of parallels between the findings of physics and mysticism, simply because we can expect some sort of similarities, however meager, between levels 1 and 5. But these similarities are rather trivial when compared with the profound differences between these dimensions of Being (as I will explain in a moment), and, further, overemphasizing these parallels invites a total confusion of the two object-domains in question.

The parallels themselves—to judge from popular expositions— usually boil down to some sort of statement about "all things being mutually interrelated in a holistic way." But if that statement is not outright wrong, it is still trivial. Personally, I believe it to be wrong: all things are not mutually or symmetrically or equivalently interrelated; in the realm of manifestation, hierarchical and asymmetrical relationships are, as we have seen, at least as important as mutual or equivalent relationships. In the realm of time, for instance, the past has affected the present but the present has no effect on the past (e.g., what Columbus did most definitely affects you, but what you do has no effect on Columbus; there is nothing mutual in that relationship at all).

But even supposing that statement is true, which it isn't, it is still trivial, for it tells us nothing the old physics couldn't tell us. According to Newtonian physics, everything in the universe was related to everything else by instantaneous action-at-a-distance, a holistic concept if ever there was one. (Incidentally, there is an excellent book on the new physics—Heinz Pagels's *The Cosmic Code*[21]—which is the only book I can unreservedly recommend on the topic. In addition to a superb explanation and discussion of the new physics, it points out—correctly I believe—that Newtonian physics is actually much closer in many ways to

Eastern mysticism than is quantum physics.) I could go on in this fashion, examining each of the supposed parallels between the findings of the new physics and those of mysticism, but the conclusion would be the same: where the alleged parallels are not simply the result of over-generalizations or foggy semantic conclusions, they are either downright wrong or trivial.

And if, finally, by mysticism we mean a direct knowledge of and as Spirit (or Ground), there are no parallels whatsoever between physics (old or new) and mysticism, for the simple reason that Spirit as Ground has no qualities with which it can be compared, contrasted, or paralleled. In order to compare Spirit with, say, the findings of physics, Spirit has to be assigned *some sort* of qualifications or set-apart characteristics, at which point it ceases absolutely to be Spirit.

But aren't physics and mysticism simply two different approaches to the same underlying Reality?

No, no, yes, and no. If by "Reality" you mean spirit (or level 5), then physics and mysticism are not dealing with the same reality at all, but with two very different levels or dimensions of reality, a confusion of which is wholly unwarranted.

If by "Reality" you mean Spirit as Ground, then no valid comparisons can be stated at all, and only Wittgenstein's commandment remains: "Whereof one cannot speak, thereof one must be silent."

If by "underlying Reality" you mean the Totality of everything that is, then obviously physics and mysticism are parts or aspects of that Totality; all you have done is invent a trivial tautology. I am rather perversely in favor of the shock value that that tautology has on orthodox scientists, but unfortunately (and quite correctly), when they investigate it more closely, they find only bogus scientific claims supporting allegedly mystical claims, which, in the long run, helps neither genuine science nor genuine mysticism.

Finally, if by "one underlying Reality" you explicitly mean Spirit, then you are attributing the particular quality of "oneness" to Spirit, which is exactly, as we have seen, the way *not* to think about Spirit. And yet it is usually that attribution that is at the heart of the considerable success of the popular "physics/mysticism" books. When Charles II was asked to explain the popularity of a rather obscure preacher, he replied, "I suppose his nonsense suits their nonsense."

Before we leave this topic, let me give a concrete example to make my somewhat difficult semantic distinctions as clear as possible. We have seen that each level in the Great Nest transcends but includes its predecessor such that level 1 could be represented as A, level 2 as A plus B, level 3 as A plus B plus C, and so on. There are more significant parallels between level 1 and 2, or 2 and 3, or 3 and 4, than there are between, say, 1 and 4 or 1 and 5—simply because the former are "closer" in terms of structural similarities and number of shared characteristics. But my point: physics has combed the material realm (level 1) and found four—and only four—major "forces": gravitational, electromagnetic, strong nuclear, and weak nuclear. By the time we reach level 2, or biological systems, we still have those four forces in effect, but we have *added* the forces of food-desire, sex-desire, water-desire, motor capacity, plus other rather elementary drives commonly called instincts. As we move to the psychological (level 3), we *add* the forces or motivations of jealousy, hope, envy, pride, guilt, remorse, justice, artistic endeavor, morality—to name a very few. And in the spiritual realms (4 and 5), we have the added forces of universal love, compassion, grace, skillful means, radical intuition, the ten *paramitas*—to also name a very few.

Now there is a legitimate type of endeavor that attempts to isolate certain commonalities between *all* these forces, but already you can see how extremely careful we must be in this endeavor. After all, there are (as far as we know) only four forces operating on level 1; by the time we reach level 5, we have added hundreds, perhaps thousands, of new and different operative forces, and whatever parallels we find between the four physical forces and the hundreds or thousands of higher forces will obviously be of the most meager variety imaginable. *I am absolutely in favor of that endeavor;* it is simply that every effort I have seen in this direction (including General System Theory) turns out to be either wrong (i.e., based on category errors) or trivial (i.e., so noncommitally abstract), and personally, I suspect that most of the genuine parallels (we call them "analog laws") will be, as I also said, rather meager.

And so that, to summarize, is why virtually all of the physicists in this volume concluded that, no matter how you slice the ontological pie, the findings of modern physics and mysticism have very little in common, other than the trivial tautology that they are all different aspects of the same reality.

But—and I should like to end on this note—every physicist in this volume was completely in favor of interdisciplinary dialogue. After intensively studying all their works for this anthology, I personally believe they would disagree with virtually all of the popular books on "physics-and-mysticism," but they would wholeheartedly applaud and support those efforts to come to terms with, we might say, the fundamental quantum questions of existence. The individuals in this volume were physicists, but they were also philosophers and mystics, and they couldn't help but muse on how the findings of physics might fit into a larger or overall worldview. I would estimate that, despite the fact (or actually because of the fact) that their common conclusion was that the domains of physics and mysticism have little or nothing in common, nonetheless, ninety percent of this volume contains ideas and opinions generated precisely by the dialogue between these two extreme limits of the Great Nest of Being. All of that, it is my own belief, is exactly as it should be. Their aim was to find physics *compatible* with a larger or mystical worldview—not confirming and not proving, but simply not contradicting. All of them, in their own ways, achieved considerable success.

Settle back now for some of the finest dialogues between physics and philosophy and religion ever authored by the human spirit.

> In the Preface I said that the attempt to "prove" mysticism with modern physics is not only wrong but actually detrimental to genuine mysticism. Now the attempt itself is perfectly understandable—those who have had a direct glimpse of the mystical know how real and how profound it is. But it is so hard to convince skeptics of this fact, that it is extremely tempting and appealing to be able to claim that physics—the "really real" science—actually supports mysticism. I, in my earliest writings, did exactly that. But it is an error, and it is detrimental, meaning, in the long run it causes much more harm than good, and for the following reasons: 1) It confuses temporal, relative, finite truth with eternal-absolute truth. Fritjof Capra has, I believe, considerably modified his views, but in *The Tao of Physics*, for instance, he put much stake in bootstrap theory (which says there are no irreducible things, only self-consistent relationships) and equated this with the Buddhist mystical doctrine of mutual interpenetration. But nowadays virtually all physicists believe there are irreducible things (quarks, leptons, gluons) that arise out of broken symmetries. Does Buddha therefore lose his enlightenment? To repeat Bernstein, "To hitch a religious philosophy to a contemporary science is a sure route to its obsolescence." And

that is detrimental. 2) It encourages the belief that in order to achieve mystical awareness all one need do is learn a new world-view; since physics and mysticism are simply two different approaches to the same reality, why bother with years of arduous meditation? Just read *The Tao of Physics*. This was obviously not Capra's intent, merely one of the unforeseen effects. 3) In the greatest irony of all, this whole approach is profoundly reductionistic. It says, in effect: since all things are ultimately made of subatomic particles, and since subatomic particles are mutually interrelated and holistic, then all things are holistically one, just like mysticism says. But all things are not ultimately made of subatomic particles; all things, including subatomic particles, are ultimately made of God. And the material realm, far from being the most significant, is the least significant: it has less Being than life, which has less Being than mind, which has less Being than soul, which has less Being than spirit. Physics is simply the study of the realm of least-Being. Claiming that all things are ultimately made of subatomic particles is thus the most reductionistic stance imaginable! I said this is ironic, because it is exactly the opposite of the obviously good intent of these new-age writers, who are trying to help mysticism while in fact they have just sunk it. The extreme (but often subtle and hidden) reductionism of this view horrifies even orthodox philosophers and scientists. Stephen Jay Gould, for instance, in a very thoughtful and sympathetic review of Capra's *The Turning Point*, finally stood back aghast: "Consider the peculiarity of that last sentence: 'the subatomic particles—and therefore, ultimately, all parts of the universe. . . .' The self-styled holist and antireductionist [i.e., Capra] is finally caught in his own parochialism after all. He has followed the oldest of the reductionist strategies. As it is with the structure of physics, so it must be, by extrapolation, with all of nature. You don't exit from this [reductionistic] trap by advocating holism at the lowest level. The very assertion that this lowest level, whatever its nature, represents the essence of reality, is the ultimate reductionist argument." Gould then goes on to point out that modern biology, psychology, and sociology work with "entities in a sequence of levels with unique explanatory principles emerging at each more inclusive plateau. This hierarchical perspective must take seriously the principle that phenomena of one level cannot automatically be extrapolated to work in the same way as others." In other words, modern orthodox scientists and philosophers are simply rediscovering the Great Chain of Being! But it is embarrassing, to say the least, for them to have to point out the blatant reductionism in the new-age "antireductionists." And this is *detrimental,* as I said, because it further alienates and polarizes the orthodox theorists who, I believe, really want to be open to hierarchical-transcendentalism, if it is presented carefully, rationally, and nonreductionistically. As it is now, most new-age approaches simply irritate the orthodox, not because these approaches are mystical but, to the contrary, because they are so reductionistic! Thus Gould, who started out his review of *The Turning Point* by saying that "This enormously right-minded general theme surely wins my approval," ended it with: "I found myself getting more and more annoyed with his book, with its facile analogies, its distrust of reason, its invocation of fashionable notions. In some respects, I feel closer to rational Cartesians [he despises them] than to Capra's California brand of ecology." (*New York Review of Books,* March 3, 1983.) I think Gould is too harsh on Capra; my point is that Capra is one of the most careful of the new-age writers, and yet even his approach is reductionistic enough to shock poor Gould into apoplexy. And the attempt continues: Arthur Young thinks absolute spirit is a photon. But wait! French physicist Jean Charon, in *The Unknown Spirit,* has just demonstrated that spirit is an electron! (I'm serious.) And, as I now write this, *God and the New Physics* has just been released. . . .

REFERENCES

1. Interview contained in M. Planck, *Where Is Science Going?* (New York: Norton, 1932), p. 209.
2. Sir Arthur Stanley Eddington, *The Nature of the Physical World* (New York: Macmillan, 1929).
3. Sir Arthur Stanley Eddington, *New Pathways in Science,* (New York: Macmillan, 1935), pp. 307–8.
4. Erwin Schroedinger, *Science, Theory, and Man* (New York: Dover, 1957), p. 204.
5. Erwin Schroedinger, *Nature and the Greeks* (Cambridge University Press, 1954), p. 8.

6. Quoted in W. Heisenberg, *Physics and Beyond* (New York: Harper and Row, 1971), pp. 82–3.
7. *Living Philosophies*, p. 117.
8. Niels Bohr, *Atomic Physics and Human Knowledge* (New York: Wiley, 1958), p. 74.
9. Quoted in W. Heisenberg, *Physics and Beyond*, p. 88.
10. Louis de Broglie, *Matter and Light* (New York: Dover, 1946), p. 252.
11. E. Schroedinger, *Nature and the Greeks*, p. 15.
12. N. Bohr, *Atomic Theory and the Description of Nature* (Cambridge University Press, 1961), p. 77.
13. Sir James Jeans, *Physics and Philosophy*, pp. 15–17.
14. A. Eddington, *The Nature of the Physical World*, p. 282.
15. Ibid.
16. E. Schroedinger, *Mind and Matter* (Cambridge University Press, 1958).
17. Sir James Jeans, *The Mysterious Universe* (Cambridge University Press, 1931), p. 111.
18. A. Eddington, *Science and the Unseen World* (New York: Macmillan, 1929).
19. I have dealt with all this in greater detail; see K. Wilber, *Eye to Eye* (New York: Doubleday/Anchor, 1983).
20. Ibid.
21. H. Pagels, *The Cosmic Code* (New York: Bantam, 1982).

HEISENBERG

WERNER HEISENBERG
(1901–1976)

IN THE SUMMER of 1925, suffering from a bout of hay fever and exhausted from wrestling with the perplexities of atomic spectral lines, Werner Heisenberg—then only twenty-four years old—took a short vacation from the Physics Institute at Gottingen University, where he was studying with Max Born, and traveled to the hills of Helgoland. There, in one fevered day and night, he invented what was to be known as matrix quantum mechanics. With the help of Max Born, Pascual Jordan, Paul Dirac, and Wolfgang Pauli, matrix quantum mechanics was formalized (one of the results of which was the famous Heisenberg Uncertainty Principle, which, in plain language, says that the more we know about half of the subatomic world, the less we can know about the other half). Erwin Schroedinger, working independently and along different lines, developed a wave mechanics; these two formalisms were quickly shown to be equivalent, and, almost at one stroke, modern quantum mechanics was born. In 1932 Heisenberg was awarded the Nobel Prize in Physics for his crucial and brilliant contributions.

The following sections are taken from *Physics and Beyond* (New York: Harper and Row, 1971), *Across the Frontiers* (New York: Harper and Row, 1974), and *The Physicist's Conception of Nature* (New York: Harcourt and Brace, 1955). His central point is that physics can make only statements "about strictly limited relations that *are only valid within the framework of those limitations* [his italics]." If we want to go beyond physics, however, and begin to philosophize, then the worldview that can most easily *explain* modern physics is that not of Democritus, but of Plato. Heisenberg was an excellent philosopher (probably, with Eddington, the most accomplished in this volume), and a metaphysician or mystic of the Pythagorean-Platonic variety. Capable of being rigorously analytical and empirical, he nonetheless despised mere positivism—or the attempt to be *only* analytical and empirical—and thus in the opening section, Heisenberg, Pauli, and Bohr lament the attempt of philosophy to ape physics.

2

Truth Dwells in the Deeps

THE RESUMPTION of international contacts once again brought together old friends. Thus, in the early summer of 1952, atomic physicists assembled in Copenhagen to discuss the construction of a European accelerator. I was most interested in this project because I was hoping that a large accelerator would help us to determine whether or not the high-energy collision of two elementary particles could lead to the production of a host of further particles, as I had assumed; whether, indeed, we were entitled to assume the existence of many new particles and, if so, whether, like the stationary states of atoms or molecules, they differed only in their symmetries, masses, and lifetimes. The main topic of the meeting was thus a matter of great personal concern, and if I do not report it here, it is simply because I must relate a conversation with Niels [Bohr] and Wolfgang [Pauli] on that occasion. Wolfgang had come over from Zurich, and the three of us were sitting in the small conservatory that ran from Bohr's official residence down to the park. We were discussing the old theme, namely, whether our interpretation of quantum theory in this very spot, twenty-five years ago, had been correct, and whether or not our ideas had since become part of the intellectual stock-in-trade of all physicists. Niels had this to say:

"Some time ago there was a meeting of philosophers, most of them positivists, here in Copenhagen, during which members of the Vienna Circle played a prominent part. I was asked to address them on the interpretation of quantum theory. After my lecture, no one raised any objections or asked any embarrassing questions, but I must say this very fact proved a terrible disappointment to me. For those who are not shocked when they first come across quantum theory cannot possibly

33

have understood it. Probably I spoke so badly that no one knew what I was talking about."

Wolfgang objected: "The fault need not necessarily have been yours. It is part and parcel of the positivist creed that facts must be taken for granted, sight unseen, so to speak. As far as I remember, Wittgenstein says: 'The world is everything that is the case.' 'The world is the totality of facts, not of things.' Now if you start from that premise, you are bound to welcome any theory representative of the 'case.' The positivists have gathered that quantum mechanics describes atomic phenomena correctly, and so they have no cause for complaint. What else we have had to add—complementarity, interference of probabilities, uncertainty relations, separation of subject and object, etc.—strikes them as just so many embellishments, mere relapses into prescientific thought, bits of idle chatter that do not have to be taken seriously. Perhaps this attitude is logically defensible, but, if it is, I for one can no longer tell what we mean when we say we have understood nature."

Niels [commented]: "For my part, I can readily agree with the positivists about the things they want, but not about the things they reject. All the positivists are trying to do is to provide the procedures of modern science with a philosophical basis, or, if you like, a justification. They point out that the notions of the earlier philosophies lack the precision of scientific concepts, and they think that any of the questions posed and discussed by conventional philosophers have no meaning at all, that they are pseudo problems and, as such, best ignored. Positivist insistence on conceptual clarity is, of course, something I fully endorse, but their prohibition of any discussion of the wider issues, simply because we lack clear-cut enough concepts in this realm, does not seem very useful to me—this same ban would prevent our understanding of quantum theory."

"Positivists," I tried to point out, "are extraordinarily prickly about all problems having what they call a prescientific character. I remember a book by Philipp Frank on causality, in which he dismisses a whole series of problems and formulations on the ground that all of them are relics of the old metaphysics, vestiges from the period of prescientific or animistic thought. For instance, he rejects the biological concepts of 'wholeness' and 'entelechy' as prescientific ideas and tries to prove that all statements in which these concepts are commonly used have no verifiable meaning. To him 'metaphysics' is a synonym for 'loose thinking,' and hence a term of abuse."

"This sort of restriction of language doesn't seem very useful to me either," Niels said. "You all know Schiller's poem, 'The Sentences of Confucius,' which contains these memorable lines: 'The full mind is alone the clear, and truth dwells in the deeps.' The full mind, in our case, is not only an abundance of experience but also an abundance of concepts by means of which we can speak about our problems and about phenomena in general. Only by using a whole variety of concepts when discussing the strange relationship between the formal laws of quantum theory and the observed phenomena, by lighting this relationship up from all sides and bringing out its apparent contradictions, can we hope to effect that change in our thought processes which is a *sine qua non* of any true understanding of quantum theory.

"You mentioned Philipp Frank's book on causality. Philipp Frank was one of the philosophers to attend the congress in Copenhagen, and he gave a lecture in which he used the term 'metaphysics' simply as a swearword or, at best, as a euphemism for unscientific thought. After he had finished, I had to explain my own position, and this I did roughly as follows:

"I began by pointing out that I could see no reason why the prefix 'meta' should be reserved for logic and mathematics—Frank had spoken of metalogic and metamathematics—and why it was anathema in physics. The prefix, after all, merely suggests that we are asking further questions, i.e., questions bearing on the fundamental concepts of a particular discipline, and why ever should we not be able to ask such questions in physics? But I should start from the opposite end. Take the question 'What is an expert?' Many people will tell you that an expert is someone who knows a great deal about his subject. To this I would object that no one can ever know very much about any subject. I would much prefer the following definition: an expert is someone who knows some of the worst mistakes that can be made in his subject, and how to avoid them. Hence Philipp Frank ought to be called an expert on metaphysics, one who knows how to avoid some of its worst mistakes—I was not quite sure whether Frank was very happy about my praise, though I was certainly not offering it tongue-in-cheek. In all such discussions what matters most to me is that we do not simply talk the 'deeps in which the truth dwells' out of existence. That would mean taking a very superficial view."

That same evening Wolfgang and I continued the discussion alone. It was the season of the long nights. The air was balmy, twilight lasted until almost midnight, and as the sun traveled just beneath the horizon, it bathed the city in a subdued, bluish light. And so we decided to walk along the Langelinie, a beautiful harbor promenade, with freighters discharging their cargo on either side. In the south, the Langelinie begins roughly where Hans Christian Andersen's Little Mermaid rests on a rock beside the beach; in the north, it is continued by a jetty that swings out into the basin and marks the entrance to Frihavn with a small beacon. After we had been looking at the toing and froing of the ships in the twilight for quite a while, Wolfgang asked me:

"Were you quite satisfied with Niels' remarks about the positivists? I gained the impression that you are even more critical of them than Niels himself, or rather that your criterion of truth differs radically from theirs."

"I should consider it utterly absurd—and Niels, for one, would agree—were I to close my mind to the problems and ideas of earlier philosophers simply because they cannot be expressed in a more precise language. True, I often have great difficulty in grasping what these ideas are meant to convey, but when that happens, I always try to translate them into modern terminology and to discover whether they throw up fresh answers. But I have no principled objections to the re-examination of old questions, much as I have no objections to using the language of any of the old religions. We know that religions speak in images and parables and that these can never fully correspond to the meanings they are trying to express. But I believe that, in the final analysis, all the old religions try to express the same contents, the same relations, and all of these hinge around questions about values. The positivists may be right in thinking that it is difficult nowadays to assign a meaning to such parables. Nevertheless, we ought to make every effort to grasp their meaning, since it quite obviously refers to a crucial aspect of reality; or perhaps we ought to try putting it into modern language, if it can no longer be contained in the old."

"If you think about such problems in that way, then, quite obviously, you cannot accept the equation of truth and predictive power. But what is your own criterion of truth in science?"

"We may find it more helpful to revert to our old comparison between Ptolemy's astronomy and Newton's conception of planetary motions. If predictive power were indeed the only criterion of truth, Ptolemy's astronomy would be no worse than Newton's. But if we compare Newton and Ptolemy in retrospect, we gain the clear impression that Newton's equations express the paths of the planets much more fully and correctly than Ptolemy's did, that Newton, so to speak, described the plan of nature's own construction. Or, to take an example from modern physics: when we learn that the principles of conservation of energy, charge, etc., have a quite universal character, that they apply in all branches of physics and that they result from the symmetry inherent in the fundamental laws, then we are tempted to say that symmetry is a decisive element in the plan on which nature has been created. In saying this I am fully aware that the words 'plan' and 'created' are once again taken from the realm of human experience and that they are metaphors at best. But it is quite easy to see that everyday language must necessarily fall short here. I suppose that is all I can say about my own conception of scientific truth."

"Quite so, but positivists will object that you are making obscure and meaningless noises, whereas they themselves are models of analytic clarity. But where must we seek for the truth, in obscurity or in clarity? Niels has quoted Schiller's 'Truth dwells in the deeps.' Are there such deeps and is there any truth? And may these deeps perhaps hold the meaning of life and death?"

A few hundred yards away, a large liner was gliding past, and its bright lights looked quite fabulous and unreal in the bright blue dusk. For a few moments, I speculated about the human destinies being played out behind the lit-up cabin windows, and suddenly Wolfgang's questions got mixed up with it all. What precisely was this steamer? Was it a mass of iron with a central power station and electric lights? Was it the expression of human intentions, a form resulting from interhuman relations? Or was it a consequence of biological laws, exerting their formative powers not merely on protein molecules but also on steel and electric currents? Did the word "intention" reflect the existence merely of these formative powers or of these biological laws in the human consciousness? And what did the word "merely" mean in this context?

My silent soliloquy now turned to more general questions. Was it utterly absurd to seek behind the ordering structures of this world a "consciousness" whose "intentions" were these very structures? Of course, even to put this question was an anthropomorphic lapse, since the word "consciousness" was, after all, based purely on human experience, and ought therefore to be restricted to the human realm. But in that case we would also be wrong to speak of animal consciousness, when we have a strong feeling that we can do so significantly. We sense that the meaning of "consciousness" becomes wider and at the same time vaguer if we try to apply it outside the human realm.

The positivists have a simple solution: the world must be divided into that which we can say clearly and the rest, which we had better pass over in silence. But can anyone conceive of a more pointless philosophy, seeing that what we can say clearly amounts to next to nothing? If we omitted all that is unclear, we would probably be left with completely uninteresting and trivial tautologies.

We walked on in silence and had soon reached the northern tip of the Langelinie, whence we continued along the jetty as far as the small beacon. In the north, we could still see a bright strip of red; in these latitudes the sun does not travel far beneath the horizon. The outlines of the harbor installations stood out sharply, and after we had been standing at the end of the jetty for a while, Wolfgang asked me quite unexpectedly:

"Do you believe in a personal God? I know, of course, how difficult it is to attach a clear meaning to this question, but you can probably appreciate its general purport."

"May I rephrase your question?" I asked. "I myself should prefer the following formulation: Can you, or anyone else, reach the central order of things or events, whose existence seems beyond doubt, as directly as you can reach the soul of another human being? I am using the term 'soul' quite deliberately so as not to be misunderstood. If you put your question like that, I would say yes. And because my own experiences do not matter so much, I might go on to remind you of Pascal's famous text, the one he kept sewn in his jacket. It was headed 'Fire' and began with the words: 'God of Abraham, Isaac and Jacob—not of the philosophers and sages.' "

"In other words, you think that you can become aware of the central order with the same intensity as of the soul of another person?"

"Perhaps."

"Why did you use the word 'soul' and not simply speak of another person?"

"Precisely because the word 'soul' refers to the central order, to the inner core of a being whose outer manifestations may be highly diverse and pass our understanding.

"If the magnetic force that has guided this particular compass—and what else was its source but the central order?—should ever become extinguished, terrible things may happen to mankind, far more terrible even than concentration camps and atom bombs. But we did not set out to look into such dark recesses; let's hope the central realm will light our way again, perhaps in quite unsuspected ways. As far as science is concerned, however, Niels is certainly right to underwrite the demands of pragmatists and positivists for meticulous attention to detail and for semantic clarity. It is only in respect to its taboos that we can object to positivism, for if we may no longer speak or even think about the wider connections, we are without a compass and hence in danger of losing our way."

Despite the late hour, a small boat made fast on the jetty and took us back to Kongens Nytorv, whence it was easy to reach Niels' house.

3

Scientific and Religious Truths

I N THE HISTORY OF SCIENCE, ever since the famous trial of Galileo, it has repeatedly been claimed that scientific truth cannot be reconciled with the religious interpretation of the world. Although I am now convinced that scientific truth is unassailable in its own field, I have never found it possible to dismiss the content of religious thinking as simply part of an outmoded phase in the consciousness of mankind, a part we shall have to give up from now on. Thus in the course of my life I have repeatedly been compelled to ponder on the relationship of these two regions of thought, for I have never been able to doubt the reality of that to which they point. In what follows, then, we shall first of all deal with the unassailability and value of scientific truth, and then with the much wider field of religion; finally—and this will be the hardest part to formulate—we shall speak of the relationship of the two truths.

Of the beginnings of modern science, the discoveries of Copernicus, Galileo, Kepler, and Newton, it is usually said that the truth of religious revelation, laid down in the Bible and the writings of the Church Fathers and dominant in the thought of the Middle Ages, was at that time supplemented by the reality of sensory experience, which could be checked by anyone in possession of his normal five senses and which—if enough care was taken—could, therefore, not in the end be doubted. But even this first approach to a description of the new way of thought is only half correct; it neglects decisive features without which its power cannot be understood. It is certainly no accident that the beginnings of modern science were associated with a turning away from Aristotle and a reversion to Plato. Even in antiquity, Aristotle, as an empiricist, had raised the objection—I cite more or less his own words—that the Pythagoreans (among whom Plato must be included) did not seek for explanations and theories to suit the facts, but distorted the facts to fit certain theories and favored opinions, and set themselves us, one might say, as coarrangers of the universe. In fact, the new science led away from immediate experience in the manner criticized by Aristotle. Let us consider the understanding of the planetary motions. Immediate experience teaches that the earth is at rest and that the sun goes around it. In the more precise terms of our own day, we might even say that the word "rest" is defined by the statement that the earth is at rest, and that we call every body at rest that no longer moves relative to the earth. If the word "rest" is understood in this fashion—and it generally *is* so understood—then Ptolemy was right and Copernicus wrong. Only if we reflect upon the concepts of "motion" and "rest," and realize that motion implies a statement about the relation between at least two bodies, can we reverse the relationship, making the sun the still center of the planetary system, and thereby obtaining a far simpler and more unified picture of the solar system, whose explanatory power was later fully recognized by Newton. Copernicus has thus appended to immediate experience a wholly new element, which I shall describe at this point as the "simplicity of natural laws," and which, in any event, has nothing to do with immediate experience. The same can be seen in Galileo's laws of falling bodies. But motion in a vacuum was at that time still quite impossible to observe. The place of immediate experience, has therefore been taken by an idealization of experience, which claims to be recognized as the correct idealization by virtue of the fact that it allows mathematical structures to become visible in the phenomena. There can be no doubt that in this early phase of modern science the newly discovered conformity to mathematical law has become the true basis for its persuasive power. These mathematical laws, so we read in Kepler, are the visible expression of the divine will, and Kepler breaks into enthusiasm at the fact that he has been the first here to recognize the beauty of God's works. Thus the new way of thinking assuredly had nothing to do with any turn away from religion. If the new discoveries did in fact contradict the teachings of the Church at certain points, this could have little significance, seeing that it was possible to perceive with such immediacy the workings of God in nature.

The God here referred to is, however, an ordering God, of whom we do not at once know whether He is identical with the God to whom we turn in trouble, and to whom we can relate our life. It may therefore be said, perhaps, that here attention was directed entirely to one aspect of the divine activity, and that hence there arose the danger of losing sight of the totality, the interconnected unity of the whole; attention is too much drawn to the narrow field of material welfare, and the other foundations of our existence are neglected. Even if technology and science could be employed merely as means to an end, the outcome depends upon whether goals for whose attainment they are to be used are good ones. But the decision upon goals cannot be made within science and technology; it is made, if we are not to go wholly astray, at a point where our vision is directed upon the whole man and the whole of his reality, not merely on a small segment of this. But this total reality contains much of which we have not said anything yet.

First, there is the fact that man can develop his mental and spiritual powers only in relation to a human society. The very capacities that distinguish him above all other living creatures, the ability to reach beyond the immediate sensory given, the recognition of wider interrelations, depend upon his being lodged in a community of speaking and thinking beings. History teaches that such communities have acquired in their development not only an outward but also a spiritual pattern. And in the spiritual patterns known to us, the relation to a meaningful connection of the whole, beyond what can be immediately seen and experienced, has almost always played the deciding role. It is only within this spiritual pattern, of the ethos prevailing in the community, that man acquires the points of view whereby he can also shape his own conduct wherever it involves more than a mere reaction to external situations; it is here that the question about values is first decided. Not only ethics, however, but the whole cultural life of the community is governed by this spiritual pattern. Only within its sphere does the close connection first become visible between the good, the beautiful, and the true, and here only does it first become possible to speak of life having a meaning for the individual. This spiritual pattern we call the religion of the community. The word "religion" is thereby endowed with a rather more general meaning than is customary. It is intended to cover the spiritual content of many cultures and different periods, even in places where the very idea of God is absent. Only in the communal modes of thought pursued in modern totalitarian states, in which the transcendent is completely excluded, would it be possible to doubt whether the concept of religion can still be meaningfully applied.

At this point, we also recognize the characteristic difference between genuine religions, in which the spiritual realm, the central spiritual order of things, plays a decisive part, and the narrower forms of thought, especially in our own day, which relate only to the strictly experienceable pattern of a human community. Such forms of thought exist in the liberal democracies of the West no less than in the totalitarian states of the East. Here, too, to be sure, an ethic is formulated, but the talk is of a norm of ethical behavior, and this norm is derived from a world outlook, that is, from inspection of the immediately visible world of experience. Religion proper speaks not of norms, however, but of guiding ideals, by which we should govern our conduct and which we can at best only approximate. These ideals do not spring from inspection of the immediately visible world but from the region of the structures lying behind it, which Plato spoke of as the world of Ideas, and concerning which we are told in the Bible, "God is a spirit."

But all that has here been said about religion is naturally well known; it has been repeated only in order to emphasize that *even the natural scientist must recognize this comprehensive significance of religion in human society,* if he wants to try to think about the relation of religious and scientific truth.

I have already sought to enunciate the thesis that in the images and likenesses of religion, we are dealing with a sort of language that makes possible an understanding of that interconnection of the world which can be traced behind the phenomena and without which we could have no ethics or scale of values. This language is in principle replaceable, like any other; in other parts of the world there are and have been other languages that provide for the same understanding. But we are born into a particular linguistic area. This language is closer akin to that of poetry than to the precision-orientated language of natural science. Hence the words in the two languages often have different meanings. The heavens referred to in the Bible have little to do with the heavens into which we send up aircraft and rockets. In the astronomical universe, the earth is only a minute grain of dust in one of the countless galactic systems, but for us it is the center of the universe—it really is the center. Science tries to give its concepts an objective meaning. But religious language must avoid this very cleavage of the world into its objective and its subjective sides; for who would dare claim the objective side to be more real than the subjective? *Thus we ought not to intermingle the two languages;* we should think more subtly than we have hitherto been accustomed to do.

The care to be taken in keeping the two languages, religious and scientific, apart from one another, should also include an *avoidance of any weakening of their content by blending them.* The correctness of tested scientific results cannot rationally be cast in doubt by religious thinking, and conversely, the ethical demands stemming from the heart of religious thinking ought not to be weakened by all too rational arguments from the field of science. There can be no doubt, in this connection, that through the enlargement of technical possibilities new ethical problems have also appeared that cannot be easily resolved. I may mention as examples the problem of the researcher's responsibility for the practical application of his discoveries, or the still more difficult question from the field of modern medicine of how long a doctor should or may prolong the life of a dying patient. Consideration of such problems has nothing to do with any watering down of ethical principles. Nor am I able to conceive that such questions are capable of being answered by pragmatic considerations of expediency alone. On the contrary, here too it will be necessary to take into account the connection of the whole—the source of ethical principles in that basic human attitude which is expressed in the language of religion.

Today, moreover, we may already be able to effect a more correct distribution of the emphases that have been misplaced by the enormous expansion of science and technology in the past hundred years. I mean the emphases we ascribe to the material and the spiritual preconditions in the human community. The material conditions are important, and it was the duty of society to eliminate the material privation of large sections of the population, once technology and science had made it possible to do so. But now that this has been done, much unhappiness remains, and we have come to see how compellingly the individual also has need, in his self-consciousness or self-understanding, for the protection the spiritual pattern of a community can provide. It is here, perhaps, that our most important tasks now lie. If there is much unhappiness among today's student body, the reason is not material hardship, but the lack of trust that makes it too difficult for the individual to give his life a meaning. We must try to overcome the isolation which threatens the individual in a world dominated by technical expediency. Theoretical deliberations about questions of psychology or social structure will avail us little here, so long as we do not succeed in finding a way back, by direct action, to a natural balance between the spiritual and material conditions of life. It will be a matter of reanimating in daily life the values grounded in the spiritual pattern of the community, of endowing them with such brilliance that the life of the individual is again automatically directed toward them.

But it is not my business to talk about society, for we were supposed to be discussing the relationship of scientific and religious truth. In the past hundred years, science has made very great advances. The wider regions of life, of which we speak in the language of our religion, may thereby have been neglected. We do not know whether we shall succeed in once more expressing the spiritual form of our future communities in the old religious language. A rationalistic play with words and concepts is of little assistance here; the most important preconditions are honesty and directness. But since ethics is the basis for the communal life of men, and ethics can only be derived from that fundamental human attitude which I have called the spiritual pattern of the community, we must bend all our efforts to reuniting ourselves, along with the younger generation, in a common human outlook. I am convinced that we can succeed in this if again we find the right balance between the two kinds of truth.

4

The Debate between
Plato and Democritus

I T WAS HERE IN THIS PART OF THE WORLD, on the coast of the Aegean Sea, that the philosophers Leucippus and Democritus pondered about the structure of matter, and down there in the marketplace, where twilight is now falling, that Socrates disputed about the basic difficulties in our modes of expression and Plato taught that the Idea, the form, was the truly fundamental pattern behind the phenomena. The problems first formulated in this country two and a half thousand years ago have occupied the human mind almost unceasingly ever since and have been discussed again and again in the course of history whenever new developments have altered the light in which the old lines of thought appeared.

If I endeavor today to take up some of the old problems concerning the structure of matter and the concept of natural law, it is because the development of atomic physics in our own day has radically altered our whole outlook on nature and the structure of matter. It is perhaps not an improper exaggeration to maintain that some of the old problems have quite recently found a clear and final solution. So it is permissible today to speak about this new and perhaps conclusive answer to questions that were formulated here thousands of years ago.

There is, however, yet another reason for renewing consideration of these problems. The philosophy of materialism, developed in antiquity by Leucippus and Democritus, has been the subject of many discussions since the rise of modern science in the seventeenth century and, in the form of dialectical materialism, has been one of the moving forces in the political changes of the nineteenth and twentieth centuries. If philosophical ideas about the structure of matter have been able to play such a role in human life, if in European society they have operated almost like an explosive and may yet perhaps do so in other parts of the world, it is even more important to know what our present scientific knowledge has to say about this philosophy. To put it in rather general and precise terms, we may hope that a philosophical analysis of recent scientific developments will contribute to a replacement of conflicting dogmatic opinions about the basic problems we have broached, by a sober readjustment to a new situation, which, in itself, can even now be regarded as a revolution in human life on this earth. But even aside from this influence of science upon our time, it may be of interest to compare the philosophical discussions in ancient Greece with the findings of experimental science and modern atomic physics. If I may already anticipate at this point the outcome of such a comparison; it seems that, in spite of the tremendous success that the concept of the atom has achieved in modern science, Plato was very much nearer to the truth about the structure of matter than Leucippus or Democritus. But it will doubtless be necessary to begin by repeating some of the most important arguments adduced in the ancient discussions about matter and life, being and becoming, before we can enter into the findings of modern science.

THE CONCEPT OF MATTER IN ANCIENT PHILOSOPHY

At the beginning of Greek philosophy there stood the dilemma of the "one" and the "many." We know that there is an ever-changing variety of phenomena appearing to our senses. Yet we believe that ultimately it should be possible to trace them back somehow to some one principle.

The founders of atomism, Leucippus and Democritus, tried to avoid the difficulty by assuming the atom to be eternal and indestructible, the only thing really existing. All other things exist only because they are composed of atoms. The antithesis of "being" and "non being" in the philosophy of Parmenides is here coarsened into that between the "full" and the "void." Being is not only one; it can be repeated infinitely many times. Being is indestructible, and therefore the atom, too, is indestructible. The void, the empty space between the atoms, allows for position and motion, and thus for properties of the atom, whereas by definition, as it were, pure being can have no other property than that of existence.

This latter part of the doctrine of Leucippus and Democritus is at once its strength and its weakness. On the one hand, it provides an immediate explanation of the different aggregate states of matter, such as ice, water, and steam, since the atoms may lie densely packed and in order beside each other, or be caught in disorder and irregular motion, or finally be separated at fairly large relative intervals in space. This part of the atomic hypothesis was therefore to prove exceedingly fruitful at a later stage. On the other hand, the atom becomes in this fashion a mere building block of matter; its properties, position, and motion in space turn it into something quite different from what was meant by the original concept of "being." The atoms can even have a finite extension, and here we have finally lost the only convincing argument for their indivisibility. If the atom has spatial properties, why should it not be divided? At least its indivisibility then becomes a physical, not a fundamental property. We can now again ask questions about the structure of the atom, and we run the risk of losing all the simplicity we had hoped to find among the smallest parts of matter. We get the impression, therefore, that in its original form the atomic hypothesis was not sufficiently subtle to explain what the philosophers really wished to understand: the simple element in the phenomena and in the structure of matter.

Still, the atomic hypothesis does go a large part of the way in the right direction. The whole multiplicity of diverse phenomena, the many observed properties of matter, can be reduced to the position and motion of the atoms. Properties such as smell or color or taste are not present in atoms. But their position and motion can evoke these properties indirectly. Position and motion seem to be much simpler concepts than the empirical qualities of taste, smell, or color. But then it naturally remains to ask what determines the position and motion of the atoms. The Greek philosophers did not attempt at this point to formulate a law of nature; the modern concept of natural law did not fit into their way of thought. Yet they seem to have thought of some kind of causal description or determinism, since they spoke of necessity, of cause and effect.

The intention of the atomic hypothesis had been to point the way from the "many" to the "one," to formulate the underlying principle, the material cause, by virtue of which all phenomena can be understood. The atoms could be regarded as the material cause, but only a general law determining their positions and velocities could actually play the part of the fundamental principle. However, when the Greek philosophers discussed the laws of nature, their thoughts were directed to static forms, geometrical symmetries, rather than to processes in space and time. The circular orbits of the planets, the regular geometrical solids, appeared to be the permanent structures of the world. The modern idea, that the position and velocity of the atom at a given time could be uniquely connected by a mathematical law with its position and velocity at a later time, did not fit into the pattern of thought of that era since it employs the concept of time in a manner that arose only out of the thinking of a much later epoch.

When Plato himself took up the problems raised by Leucippus and Democritus, he adopted the idea of smallest units of matter, but he took the strongest exception to the tendency of that philosophy to suppose the atoms to be the foundation of all existence, the only truly existing material objects. Plato's atoms were not strictly material, being thought of as geometrical forms, the regular solids of the mathematicians. These bodies, in keeping with the starting point of his idealistic philosophy, were in some sense the Ideas underlying the structure of matter and characterizing the physical behavior of the elements to which they belonged. The cube, for example, was the smallest particle of the element earth and thereby symbolized at the same time the earth's stability. The tetrahedron, with its sharp points, represented the smallest particle of the element fire. The icosahedron, which comes closest among the regular solids to a sphere, stood for the mobility of the element water. In this way the regular solids were able to serve as symbols for certain tendencies in the physical behavior of matter.

But they were not strictly atoms, not indivisible basic units like those of the materialist philosophy. Plato regarded them as composed from the triangles forming their surfaces; therefore, by exchanging triangles, these smallest particles could be commuted into each other. Thus two atoms of air, for example, and one of fire could be compounded into an atom of water. In this way Plato was able to escape the problem of the indefinite divisibility of matter. For as two-dimensional surfaces the triangles were not bodies, not matter any longer; hence matter could not be further divided ad infinitum. At the lower end, therefore, in the realm, that is, of minimal spatial dimensions, the concept of matter is resolved into that of mathematical form. This form determines the behavior, first of the smallest parts of matter, then of matter itself. To a certain extent it replaces the natural law of later physics; for without making explicit references to the course of time, it characterizes the tendencies in the behavior of matter. One might say, perhaps, that the fundamental tendencies were represented by the geometrical shape of the smallest units, while the finer details of those tendencies found expression in the relative position and velocity of these units.

This whole description fits exactly into the central ideas of Plato's idealist philosophy. The structure underlying the phenomena is not given by material objects like the atoms of Democritus but by the form that determines the material objects. The Ideas are more fundamental than the objects. And since the smallest parts of matter have to be the objects whereby the simplicity of the world becomes visible, whereby we approximate to the "one" and the "unity" of the world, the Ideas can be described mathematically—they are simply mathematical forms. The saying "God is a mathematician," which in this form assuredly derives from a later period of philosophy, has its origin in this passage from the Platonic philosophy.

The importance of this step in philosophical thought can hardly be reckoned too highly. It can be seen as the decisive beginning of the mathematical science of nature, and hence be made responsible also for the later technical applications that have altered the whole picture of the world. By this step it is also first established what the term "understanding" is to mean. Among all the possible forms of understanding, the one form practiced in mathematics is singled out as the "true" understanding. Whereas all language, indeed, all art and all poetry in some way mediate understanding, it is here maintained that only the employment of a precise, logically consistent language, a language so far capable of formalization that proofs become possible, can lead to true understanding. One feels the strength of the impression made upon the Greek philosophers by the persuasive force of logical and mathematical arguments. They are obviously overwhelmed by this force. But perhaps they surrendered too early at this point.

THE ANSWER OF MODERN SCIENCE TO THE OLD PROBLEMS

If we trace the history of physics from Newton to the present day, we see that, despite the interest in details, very general laws of nature have been formulated on several occasions. The nineteenth century saw an exact working out of the statistical theory of heat. The theories of electromagnetism and special relativity have proved susceptible of combination into a very general group of natural laws containing statements not only about electrical phenomena but also about the structure of space and time. In our own century, the mathematical formulation of the quantum theory has led to an understanding of the outer shells of chemical atoms, and thus of the chemical properties of matter generally. The relations and connections between these different laws, especially between relativity and quantum theory, are not yet fully explained. But the latest developments in particle physics permit one to hope that these relations may be satisfactorily analyzed in the relatively near future. We are thus already in a position to consider what answers can be given by this whole scientific development to the questions of the old philosophers.

During the nineteenth century, the development of chemistry and the theory of heat conformed very closely to the ideas first put forward by Leucippus and Democritus. A revival of the materialist philosophy in its modern form, that of dialectical materialism, was thus a natural counterpart to the impressive advances made during this period in chemistry and physics. The concept of the atom had proved exceptionally fruitful in the explanation of chemical bonding and the physical behavior of gases. It was soon found, however, that the particles called atoms by the chemists were composed of still smaller units. But these smaller units, the electrons, followed by the atomic nuclei and finally the elementary particles, protons and neutrons, also still seemed to be atoms from the standpoint of the materialist philosophy. The fact that, at least indirectly, one can actually see a single elementary particle—in a cloud chamber, say, or a bubble chamber—supports the view that the smallest units of matter are real physical objects, existing in the same sense that stones or flowers do.

But the inherent difficulties of the materialist theory of the atom, which had become apparent even in the ancient discussions about smallest particles, have also appeared very clearly in the development of physics during the present century.

This difficulty relates to the question whether the smallest units are ordinary physical objects, whether they exist in the same way as stones or flowers. Here, the development of quantum theory some forty years ago has created a complete change in the situation. The mathematically formulated laws of quantum theory show clearly that our ordinary intuitive concepts cannot be unambiguously applied to the smallest particles. All the words or concepts we use to describe ordinary physical objects, such as position, velocity, color, size, and so on, become indefinite and problematic if we try to use them of elementary particles. I cannot enter here into the details of this problem, which has been discussed so frequently in recent years. But it is important to realize that, while the behavior of the smallest particles cannot be unambiguously described in ordinary language, the language of mathematics is still adequate for a clear-cut account of what is going on.

During the coming years, the high-energy accelerators will bring to light many further interesting details about the behavior of elementary particles. But I am inclined to think that the answer just considered to the old philosophical problems will turn out to be final. If this is so, does this answer confirm the views of Democritus or Plato?

I think that on this point modern physics has definitely decided for Plato. For the smallest units of matter are, in fact, not physical objects in the ordinary sense of the word; they are forms, structures or—in Plato's sense—Ideas, which can be unambiguously spoken of only in the language of mathematics. Democritus and Plato both had hoped that in the smallest units of matter they would be approaching the "one," the unitary principle that governs the course of the world. Plato was convinced that this principle can be expressed and understood only in mathematical form. The central problem of theoretical physics nowadays is the mathematical formulation of the natural law underlying the behavior of the elementary particles. From the experimental situation we infer that a satisfactory theory of the elementary particles must at the same time be a theory of physics in general, and hence, of everything else belonging to this physics.

In this way, a program could be carried out that in modern times was first proposed by Einstein: a unified theory of matter—and hence, simultaneously, a quantum theory of matter—could be formulated, which might serve quite generally as a foundation for physics. We do not yet know whether the mathematical forms proposed for this unifying principle are already adequate or will have to be replaced by forms more abstract still. But our present knowledge of the elementary particles is certainly enough for us to say what the main content of this law has to be. It must essentially set forth a small number of fundamental symmetry properties in nature, which have been known empirically for some years; in addition to these symmetries, it must contain the principle of causality as understood in relativity theory. The most important of the symmetries are the so-called "Lorentz group" of special relativity theory, which includes the key statements about space and time, and the so-called "isospin group," which has to do with the electric charge on the elementary particles. There are also other symmetries, but of these I shall say nothing here. Relativistic causality is connected with the Lorentz group but must be considered an independent principle.

This situation reminds us at once of the symmetrical bodies introduced by Plato to represent the fundamental structures of matter. Plato's symmetries were not yet the correct ones, but he was right in believing that ultimately, at the heart of nature, among the smallest units of matter, we find mathematical symmetries. It was an unbelievable achievement of the ancient philosophers to have asked the right questions. But, lacking all knowledge of the empirical details, we could not have expected them to find answers that were correct in detail as well.

Consequences for the Evolution of Human Thought in Our Own Day

The search for the "one," for the ultimate source of all understanding, has doubtless played a similar role in the origin of both religion and science. But the scientific method that was developed in the sixteenth and seventeenth centuries, the interest in those details which can be tested by experiment, has for a long time pointed science along a different path. It is not surprising that this attitude should have led to a conflict between science and religion, as soon as a law contradicted, in some particular and perhaps very important detail, the general picture, the mode and manner, in which the facts had been spoken of in religion. This conflict, which began in modern times with the celebrated trial of Galileo, has been discussed often enough, and I need not repeat this discussion here. One may recall that, even in ancient Greece, Socrates was condemned to death because his teachings seemed to contradict the traditional religion. In the nineteenth century, this conflict reached its peak in the attempt of some philosophers to replace traditional Christianity by a scientific philosophy, based upon a materialist version of the Hegelian dialectic. It might be said that, in directing their gaze upon a materialistic interpretation of the "one," the scientists were attempting to find their way back again to this "one" from the multitude of details.

If modern science has something to contribute to this problem, it is not by deciding for or against one of these doctrines; for example, as was possibly believed in the nineteenth century, by coming down in favor of materialism and against the Christian philosophy, or, as I now believe, in favor of Plato's idealism and against the materialism of Democritus. On the contrary, the chief profit we can derive in these problems from the progress of modern science is to learn how cautious we have to be with language and with the meaning of words. I would therefore like to devote the last part of my address to a few remarks about the problem of language in modern science and ancient philosophy.

If we may take our cue at this point from Plato's dialogues, the unavoidable limitations of our means of expression were already a central theme in the philosophy of Socrates; one might even say that his whole life was a constant battle with these limitations. Socrates never wearied of explaining to his countrymen, here on the streets of Athens, that they did not know exactly what they meant by the words they were employing. The story goes that one of Socrates' opponents, a sophist who was annoyed at Socrates' constant reference to this insufficiency of language, criticized him and said: "But Socrates, this is a bore; you are always saying the same about the same." Socrates replied: "But you sophists, who are so clever, perhaps never say the same about the same."

The reason for laying such stress on this problem of language was doubtless that Socrates was aware, on the one hand, of how many misunderstandings can be engendered by a careless use of language, how important it is to use precise terms and to elucidate concepts before employing them. On the other hand, he probably also realized that this may ultimately be an insoluble task. The situation confronting us in our attempt to "understand" may drive us to conclude that our existing means of expression do not allow of a clear and unambiguous description of the facts.

The tension between the demand for complete clarity and the inevitable inadequacy of existing concepts has been especially marked in modern science. In atomic physics we make use of a highly developed mathematical language that satisfies all the requirements in regard to clarity and precision. At the same time, we recognize that we cannot describe atomic phenomena without ambiguity in any ordinary language; we cannot, for example, speak unambiguously about the behavior of an electron in the interior of an atom. It would be premature, however, to insist that we should avoid the difficulty by confining ourselves to the use of mathematical language. This is no genuine way out, since we do not know how far the mathematical language can be applied to the phenomena. In the last resort, even science must rely upon ordinary language, since it is the only language in which we can be sure of really grasping the phenomena.

This situation throws some light on the tension between the scientific method, on the one hand, and the relation of society to the "one," the fundamental principle behind the phenomena, on the other. It seems obvious that this latter relation cannot and should not be expressed in a precise and highly sophisticated language whose applicability to the real world may be very restricted. The only thing that will do for this purpose is the natural language everyone can understand. Reliable results in science, however, can be secured only by unambiguous statement; here we cannot do without the precision and clarity of an abstract mathematical language.

The necessity of constantly shuttling between the two languages is, unfortunately, a chronic source of misunderstandings, since in many cases the same words are employed in both. The difficulty is unavoidable. But it may yet be of some help always to bear in mind that modern science is obliged to make use of both languages, that the same word may have very different meanings in each of them, that different criteria of truth apply, and that one should not, therefore, talk too hastily of contradictions.

If we wish to approach the "one" in the terms of a precise scientific language, we must turn our attention to that center of science described by Plato, in which the fundamental mathematical symmetries are to be found. In the concepts of this language we must be content with the statement that "God is a mathematician"; for we have freely chosen to confine our vision to that realm of being which can be understood in the mathematical sense of the word "understanding," which can be described in rational terms.

Plato himself was not content with this restriction. Having pointed out with the utmost clarity the possibilities and limitations of precise language, he switched to the language of poetry, which evokes in the hearer images conveying understanding of an altogether different kind. I shall not seek to discuss here what this kind of understanding can really mean. These images are probably connected with the unconscious mental patterns the psychologists speak of as archetypes, forms of strongly emotional character that, in some way, reflect the internal structures of the world. But whatever the explanation for these other forms of understanding, the language of images and likenesses is probably the only way of approaching the "one" from more general domains. If the harmony in a society rests on a common interpretation of the "one," the unitary principle behind the phenomena, then the language of poetry may be more important here than the language of science.

5

Science and the Beautiful

Perhaps it will be best if, without any initial attempt at a philosophical analysis of the concept of "beauty," we simply ask where we can meet the beautiful in the sphere of exact science. Here I may perhaps be allowed to begin with a personal experience. When, as a small boy, I was attending the lowest classes of the Max-Gymnasium here in Munich, I became interested in numbers. It gave me pleasure to get to know their properties, to find out, for example, whether they were prime numbers or not, and to test whether they could perhaps be represented as sums of squares, or eventually to prove that there must be infinitely many primes. Now since my father thought my knowledge of Latin to be much more important than my numerical interests, he brought home to me one day from the National Library a treatise written in Latin by the mathematician Leopold Kronecker, in which the properties of whole numbers were set in relation to the geometrical problem of dividing a circle into a number of equal parts. How my father happened to light on this particular investigation from the middle of the last century I do not know. But the study of Kronecker's work made a deep impression on me. I sensed a quite immediate beauty in the fact that, from the problem of partitioning a circle, whose simplest cases were, of course, familiar to us in school, it was possible to learn something about the totally different sort of questions involved in elementary number theory. Far in the distance, no doubt, there already floated the question whether whole numbers and geometrical forms exist, i.e., whether they are there outside the human mind or whether they have merely been created by this mind as instruments for understanding the world. But at that time I was not yet able to think about such problems.

The impression of something very beautiful was, however, perfectly direct; it required no justification or explanation.

But what was beautiful here? Even in antiquity there were two definitions of beauty which stood in a certain opposition to one another. The controversy between them played a great part, especially during the Renaissance. The one describes beauty as the proper conformity of the parts to one another, and to the whole. The other, stemming from Plotinus, describes it, without any reference to parts, as the translucence of the eternal splendor of the "one" through the material phenomenon. In our mathematical example, we shall have to stop short, initially, at the first definition. The parts here are the properties of whole numbers and laws of geometrical constructions, while the whole is obviously the underlying system of mathematical axioms to which arithmetic and Euclidean geometry belong—the great structure of interconnection guaranteed by the consistency of the axiom system. We perceive that the individual parts fit together, that, as parts, they do indeed belong to this whole, and, without any reflection, we feel the completeness and simplicity of this axiom system to be beautiful. Beauty is therefore involved with the age-old problem of the "one" and the "many" which occupied—in close connection with the problem of "being" and "becoming"—a central position in early Greek philosophy.

Since the roots of exact science are also to be found at this very point, it will be as well to retrace in broad outline the currents of thought in that early age. At the starting point of the Greek philosophy of nature there stands the question of a basic principle, from which the colorful variety of phenomena can be explained. However strangely it may strike us, the well-known answer of Thales—"Water is the material first principle of all things"—contains, according to Nietzsche, three basic philosophical demands which were to become important in the developments that followed: first, that one should seek for such a unitary basic principle; second, that the answer should be given only rationally, that is, not by reference to a myth; and third and finally, that in this context the material aspect of the world must play a deciding role. Behind these demands there stands, of course, the unspoken recognition that understanding can never mean anything more than the perception of connections, i.e., unitary features or marks of affinity in the manifold.

But if such a unitary principle of all things exists, then—and this was the next step along this line of thought—one is straightway brought up against the question how it can serve to account for the fact of change. The difficulty is particularly apparent in the celebrated paradox of Parmenides. Only being is; non-being is not. But if only being is, there cannot be anything outside this being that articulates it or could bring about changes. Hence being will have to be conceived as eternal, uniform, and unlimited in space and time. The changes we experience can thus be only an illusion.

Greek thought could not stay with this paradox for long. The eternal flux of appearances was immediately given, and the problem was to explain it. In attempting to overcome the difficulty, various philosophers struck out in different directions. One road led to the atomic theory of Democritus. In addition to being, non-being can still exist as a possibility, namely as the possibility for movement and form, or, in other words, as empty space. Being is repeatable, and thus we arrive at the picture of atoms in the void—the picture that has since become infinitely fruitful as a foundation for natural science. But of this road we shall say no more just now. Our purpose, rather, is to present in more detail the other road, which led to Plato's Ideas, and which carried us directly into the problem of beauty.

This road begins in the school of Pythagoras. It is there that the notion is said to have originated that mathematics, the mathematical order, was the basic principle whereby the multiplicity of phenomena could be accounted for. Of Pythagoras himself we know little. His disciples seem, in fact, to have been a religious sect, and only the doctrine of transmigration and the laying down of certain moral and religious rules and prohibitions can be traced with any certainty to Pythagoras. But among these disciples—and this was what mattered subsequently—a preoccupation with music and mathematics played an important role. Here it was that Pythagoras is said to have made the famous discovery that vibrating strings under equal tension sound together in harmony if their lengths are in a simple numerical ratio. The mathematical structure, namely the numerical ratio as a source of harmony, was certainly one of the most momentous discoveries in the history of mankind. The harmonious concord of two strings yields a beautiful sound. Owing to the discomfort caused by beat-effects, the human ear finds dissonance disturbing, but consonance, the peace of harmony, it finds beautiful. Thus the mathematical relation was also the source of beauty.

Beauty, so the first of our ancient definitions ran, is the proper conformity of the parts to one another and to the whole. The parts here are the individual notes, while the whole is the harmonious sound. The mathematical relation can, therefore, assemble two initially independent parts into a whole, and so produce beauty. This discovery effected a breakthrough, in Pythagorean doctrine, to entirely new forms of thought, and so brought it about that the ultimate basis of all being was no longer envisaged as a sensory material—such as water, in Thales— but as an ideal principle of form. This was to state a basic idea which later provided the foundation for all exact science. Aristotle, in his *Metaphysics,* reports that the Pythagoreans, ". . . who were the first to take up mathematics, not only advanced this study, but also having been brought up in it they thought its principles were the principles of all things. . . . Since, again, they saw that the modifications and the ratios of the musical scales were expressible in numbers; since, then, all other things seemed in their whole nature to be modelled on numbers; and numbers seemed to be the first things in the whole of nature, they supposed the elements of numbers to be the elements of all things, and the whole heaven to be a musical scale and a number."

Understanding of the colorful multiplicity of the phenomena was thus to come about by recognizing in them unitary principles of form, which can be expressed in the language of mathematics. By this, too, a close connection was established between the intelligible and the beautiful. For if the beautiful is conceived as a conformity of the parts to one another and to the whole, and if, on the other hand, all understanding is first made possible by means of this formal connection, the experience of the beautiful becomes virtually identical with the experience of connections either understood or, at least, guessed at.

The next step along this road was taken by Plato with the formulation of his theory of Ideas. Plato contrasts the imperfect shapes of the corporeal world of the senses with the perfect forms of mathematics; the imperfectly circular orbits of the stars, say, with the perfection of the mathematically defined circle. Material things are the copies, the shadow images, of ideal shapes in reality; moreover, as we should be tempted to continue nowadays, these ideal shapes are actual because and insofar as they become "act"-ive in material events. Plato thus distinguishes here with complete clarity a corporeal being accessible to the senses and a purely ideal being apprehensible not by the senses but only through acts of mind. Nor is this ideal being in any way in need of man's thought in order to be brought forth by him. On the contrary, it is the true being, of which the corporeal world and human thinking are mere reproductions. As their name already indicates, the apprehension of Ideas by the human mind is more an artistic intuiting, a half-conscious intimation, than a knowledge conveyed by the understanding. It is a reminiscence of forms that were already implanted in this soul before its existence on earth. The central Idea is that of the Beautiful and the Good, in which the divine becomes visible and at sight of which the wings of the soul begin to grow. A passage in the *Phaedrus* expresses the following thought: the soul is awe-stricken and shudders at the sight of the beautiful, for it feels that something is evoked in it that was not imparted to it from without by the senses but has always been already laid down there in a deeply unconscious region.

But let us come back once more to understanding and thus, to natural science. The colorful multiplicity of the phenomena can be understood, according to Pythagoras and Plato, because and insofar as it is underlain by unitary principles of form susceptible of mathematical representation. This postulate already constitutes an anticipation of the entire program of contemporary exact science. It could not, however, be carried through in antiquity, since an empirical knowledge of the details of natural processes was largely lacking.

The first attempt to penetrate into these details was undertaken, as we know, in the philosophy of Aristotle. But in view of the infinite wealth initially presented here to the observing student of nature and the total lack of any sort of viewpoint from which an order might have been discernible, the unitary principles of form sought by Pythagoras and Plato were obliged to give place to the description of details. Thus there arose the conflict that has continued to this day in the debates, for example, between experimental and theoretical physics; the conflict between the empiricist, who by careful and scrupulous detailed investigation first furnishes the presuppositions for an understanding of nature, and the theoretician, who creates mathematical pictures whereby he seeks to order and so to understand nature—mathematical pictures that prove themselves, not only by their correct depiction of experience, but also and more especially by their simplicity and beauty, to be the true Ideas underlying the course of nature.

Aristotle, as an empiricist, was critical of the Pythagoreans, who, he said, "are not seeking for theories and causes to account for observed facts, but rather forcing their observations and trying to accommodate them to certain theories and opinions of their own" and were thus setting up, one might say, as joint organizers of the universe. If we look back on the history of the exact sciences, it can perhaps be asserted that the correct representation of natural phenomena has evolved from this very tension between the two opposing views. Pure mathematical speculation becomes unfruitful because from playing with the wealth of possible forms it no longer finds its way back to the small number of forms according to which nature is actually constructed. And pure empiricism becomes unfruitful because it eventually bogs down in endless tabulation without inner connection. Only from the tension, the interplay between the wealth of facts and the mathematical forms that may possibly be appropriate to them, can decisive advances spring.

But in antiquity this tension was no longer acceptable and thus, the road to knowledge diverged for a long time from the road to the beautiful. The significance of the beautiful for the understanding of nature became clearly visible again only at the beginning of the modern period, once the way back had been found from Aristotle to Plato. And only through this change of course did the full fruitfulness become apparent of the mode of thought inaugurated by Pythagoras and Plato.

This is most clearly shown in the celebrated experiments on falling bodies that Galileo probably did not, in fact, conduct from the leaning tower of Pisa. Galileo begins with careful observations, paying no attention to the authority of Aristotle, but, following the teaching of Pythagoras and Plato, he does try to find mathematical forms corresponding to the facts obtained by experiment and thus, arrives at his laws of falling bodies. However, and this is a crucial point, he is obliged, in order to recognize the beauty of mathematical forms in the phenomena, to idealize the facts, or, as Aristotle disparagingly puts it, to force them. Aristotle had taught that all moving bodies not acted upon by external forces eventually come to rest, and this was the general experience. Galileo maintains, on the contrary, that, in the absence of external forces, bodies continue in a state of uniform motion. Galileo could venture to force the facts in this way because he could point out that moving bodies are, of course, always exposed to a frictional resistance and that motion, in fact, continues the longer, the more effectively the frictional forces can be cut off. In exchange for this forcing of the facts, this idealization, he obtained a simple mathematical law, and this was the beginning of modern exact science.

Some years later, Kepler succeeded in discovering new mathematical forms in the data of his very careful observations of the planetary orbits and in formulating the three famous laws that bear his name. How close Kepler felt himself in these discoveries to the ancient arguments of Pythagoras, and how much the beauty of the connections guided him in formulating them, can be seen from the fact that he compared the revolutions of the planets about the sun with the vibrations of a string and spoke of a harmonious concord of the different planetary orbits, of a harmony of the spheres. At the end of his work on the harmony of the universe, he broke out into this cry of joy: "I thank thee, Lord God our Creator, that thou allowest me to see the beauty in thy work of creation." Kepler was profoundly struck by the fact that here he had chanced upon a central connection which had not been conceived by man, which it had been reserved to him to recognize for the first time—a connection of the highest beauty. A few decades later, Isaac Newton in England set forth this connection in all its completeness and described it in detail in his great work *Principia Mathematica*. The road of exact science was thus pointed out in advance for almost two centuries.

But are we dealing here with knowledge merely, or also with the beautiful? And if the beautiful is also involved, what role did it play in the discovery of these connections? Let us again recall the first definition given in antiquity: "Beauty is the proper conformity of the parts to one another and to the whole." That this criterion applies in the highest degree to a structure like Newtonian mechanics is something that scarcely needs explaining. The parts are the individual mechanical processes—those which we carefully isolate by means of apparatus no less than those which occur inextricably before our eyes in the colorful play of phenomena. And the whole is the unitary principle of form which all these processes comply with and which was mathematically established by Newton in a simple system of axioms. Unity and simplicity are not, indeed, precisely the same. But the fact that in such a theory the many are confronted with the one, that in it the many are unified, itself has the undoubted consequence that we also feel it at the same time to be simple and beautiful. The significance of the beautiful for the discovery of the true has at all times been recognized and emphasized. The Latin motto *"Simplex sigillum veri"*—"The simple is the seal of the true"—is inscribed in large letters in the physics auditorium of the University of Göttingen as an admonition to those who would discover what is new; another Latin motto, *"Pulchritudo splendor veritatis"*—"Beauty is the splendor of truth"—can also be interpreted to mean that the researcher first recognizes truth by this splendor, by the way it shines forth.

Twice more in the history of exact science, this shining forth of the great connection has been the crucial signal for a significant advance. I am thinking here of two events in the physics of our own century: the emergence of relativity theory and the quantum theory. In both cases, after years of vain effort at understanding, a bewildering plethora of details has been almost suddenly reduced to order by the appearance of a connection, largely unintuitable but still ultimately simple in its substance, that was immediately found convincing by virtue of its completeness and abstract beauty—convincing, that is, to all who could understand and speak such an abstract language.

But now, instead of pursuing the historical course of events any further, let us rather put the question quite directly: What is it that shines forth here? How comes it that with this shining forth of the beautiful into exact science the great connection becomes recognizable, even before it is understood in detail and before it can be rationally demonstrated? In what does the power of illumination consist, and what effect does it have on the onward progress of science?

Perhaps we should begin here by recalling a phenomenon that may be described as the unfolding of abstract structures. It can be illustrated by the example of number theory, which we referred to at the outset, but one may also point to comparable processes in the evolution of art. For the mathematical foundation of arithmetic, or the theory of numbers, a few simple axioms are sufficient, which, in fact, merely define exactly what counting is. But with these few axioms we have already posited that whole abundance of forms which has entered the minds of mathematicians only in the course of the long history of the subject—the theory of prime numbers, of quadratic residues, of numerical congruences, etc. One might say that the abstract structures posited in and with numbers have unfolded visibly only in the course of mathematical history, that they have generated the wealth of propositions and relationships that makes up the content of the complicated science of number theory. A similar position is also occupied—at the outset of an artistic style in architecture, say—by certain simple basic forms, such as the semicircle and rectangle in Romanesque architecture. From these basic forms there arise in the course of history new, more complicated, and also altered forms, which yet can still, in some way, be regarded as variations on the same theme; thus, from the basic structures there emerges a new manner, a new style of building. We have the feeling, nonetheless, that the possibilities of development were already perceivable in these original forms, even at the outset; otherwise, it would be scarcely comprehensible that many gifted artists should have so quickly resolved to pursue these new possibilities.

Such an unfolding of abstract basic structures has assuredly also occurred in the instances I have enumerated from the history of the exact sciences. This growth, this constant development of new branches, went on in Newtonian mechanics up to the middle of the last century. In relativity theory and the quantum theory we have experienced a similar development in the present century, and the growth has not yet come to an end.

Moreover, in science, as in art, this process also has an important social and ethical aspect; for many men can take an active part in it. When a great cathedral was to be built in the Middle Ages, many master masons and craftsmen were employed. They were imbued with the idea of beauty posited by the original forms and were compelled by their task to carry out exact and meticulous work in accordance with these forms. In similar fashion, during the two centuries following Newton's discovery, many mathematicians, physicists, and technicians were called upon to deal with specific mechanical problems according to the Newtonian methods, to carry out experiments, or to effect technical applications; here, too, extreme care was always required in order to attain what was possible within the framework of Newtonian mechanics. Perhaps it may be said in general that by means of the underlying structures, in this case Newtonian mechanics, guidelines were drawn or even standards of value set up whereby it could be objectively decided whether a given task had been well or ill discharged. It is the very fact that specific requirements have been laid down, that the individual can assist by small contributions in the attainment of large goals, and that the value of his contribution can be objectively determined, which gives rise to the satisfaction proceeding from such a development for the large number of people involved. Hence even the ethical significances of technology for our present age should not be underestimated.

The development of science and technology has also produced, for example, the Idea of the airplane. The individual technician who assembles some component for such a plane, the artisan who makes it, knows that his work calls for the utmost care and exactitude and that the lives of many may well depend upon its reliability. Hence he can take pride in a well-executed piece of work, and delights, as we do, in the beauty of the aircraft, when he feels that in it the technical goal has been realized by properly adequate means. Beauty, so runs the ancient definition we have already often cited, is the proper conformity of the parts to one another and to the whole, and this requirement must also be satisfied in a good aircraft.

But in pointing thus to the evolution of beauty's ground structure, to the ethical values and demands that subsequently emerge in the historical course of development, we have not yet answered the question we asked earlier, namely, what it is that shines forth in these structures, how the great connection is recognized even before it is rationally understood in detail. Here we ought to reckon in advance with the possibility that even such recognition may be founded upon illusions. But it cannot be doubted that there actually is this perfectly immediate recognition, this shuddering before the beautiful, of which Plato speaks in the *Phaedrus*.

Among all those who have pondered on this question, it seems to have been universally agreed that this immediate recognition is not a consequence of discursive (i.e., rational) thinking. I should like here to cite two statements, one from Johannes Kepler, who has already been referred to, and the other, in our own time, from the Zürich atomic physicist Wolfgang Pauli, who was a friend of the psychologist, Carl Jung. The first passage is to be found in Kepler's *Harmony of the World*:

That faculty which perceives and recognizes the noble proportions in what is given to the senses, and in other things situated outside itself, must be ascribed to the soul. It lies very close to the faculty which supplies formal schemata to the senses, or deeper still, and thus adjacent to the purely vital power of the soul, which does not think discursively, i.e., in conclusions, as the philosophers do, and employs no considered method, and is thus not peculiar only to man, but also dwells in wild animals and the dear beasts of the field. . . . Now it might be asked how this faculty of the soul, which does not engage in conceptual thinking, and can therefore have no proper knowledge of harmonic relations, should be capable of recognizing what is given in the outside world. For to recognize is to compare the sense perception outside with the original pictures inside, and to judge that it conforms to them. Proclus has expressed the matter very finely in his simile of awakening, as from a dream. For just as the sensorily presented things in the outer world recall to us those which we formerly perceived in the dream, so also the mathematical relations given in sensibility call forth those intelligible archetypes which were already given inwardly beforehand, so that they now shine forth truly and vividly in the soul, where before they were only obscurely present there. But how have they come to be within? To this I answer that all pure Ideas or archetypal patterns of harmony, such as we were speaking of, are inherently present in those who are capable of apprehending them. But they are not first received into the mind by a conceptual process, being the product, rather, of a sort of instinctive intuition of pure quantity, and are innate in these individuals, just as the number of petals in a plant, say, is innate in its form principle, or the number of its seed chambers is innate in the apple.

So far Kepler. He is, therefore, referring us here to possibilities already to be found in the animal and plant kingdoms, to innate archetypes that bring about the recognition of forms. In our own day, Adolf Portmann, in particular, has described such possibilities, pointing, for example, to specific color patterns seen in the plumage of birds, which can possess a biological meaning only if they are also perceived by other members of the same species. The perceptual capacity will therefore have to be just as innate as the pattern itself. We may also consider bird song at this point. At first, the biological requirement here may well have been simply for a specific acoustic signal, serving to seek out the partner and understood by the latter. But to the extent that this immediate biological function declines in importance, a playful enlargement of the stock of forms may ensue, an unfolding of the underlying melodic structure, which is then found enchanting as song by even so alien a species as man. The capacity to recognize this play of forms must, at all events, be innate to the species of bird in question for certainly it has no need of discursive, rational thought. In man, to cite another example, there is probably an inborn capacity for understanding certain basic forms of the language of gesture and thus, for deciding, say, whether the other has friendly or hostile intentions—a capacity of the utmost importance for man's communal life.

Ideas similar to those of Kepler have been put forward in an essay by Pauli. He writes:

> The process of understanding in nature, together with the joy that man feels in understanding, i.e., in becoming acquainted with new knowledge, seems therefore to rest upon a correspondence, a coming into congruence of preexistent internal images of the human psyche with external objects and their behavior. This view of natural knowledge goes back, of course, to Plato and was . . . also very plainly adopted by Kepler. The latter speaks, in fact, of Ideas, preexistent in the mind of God and imprinted accordingly upon the soul, as the image of God. These primal images, which the soul can perceive by means of an innate instinct, Kepler calls archetypes. There is very wide-ranging agreement here with the primordial images or archetypes intro-

duced into modern psychology by C. G. Jung, which function as instinctive patterns of ideation. At this stage, the place of clear concepts is taken by images of strongly emotional content, which are not thought but are seen pictorially, as it were, before the mind's eye. Insofar as these images are the expression of a suspected but still unknown state of affairs, they can also be called symbolic, according to the definition of a symbol proposed by Jung. As ordering operators and formatives in this world of symbolic images, the archetypes function, indeed, as the desired bridge between sense perceptions and Ideas, and are therefore also a necessary precondition for the emergence of a scientific theory. Yet one must beware of displacing this *a priori* of knowledge into consciousness, and relating it to specific, rationally formulable Ideas.

In the further course of his inquiries, Pauli then goes on to show that Kepler did not derive his conviction of the correctness of the Copernican system primarily from any particular data of astronomical observation, but rather from the agreement of the Copernican picture with an archetype which Jung calls a *mandala* and which was also used by Kepler as a symbol for the Trinity. God, as prime mover, is seen at the center of a sphere; the world, in which the Son works, is compared with the sphere's surface; the Holy Ghost corresponds to the beams that radiate from center to surface of the sphere. It is naturally characteristic of these primal images that they cannot really be rationally or even intuitively described.

Although Kepler may have acquired his conviction of the correctness of Copernicanism from primal images of this kind, it remains a crucial precondition for any usable scientific theory that it should subsequently stand up to empirical testing and rational analysis. In this respect, the sciences are in a happier position than the arts, since for science there is an inexorable and irrevocable criterion of value that no piece of work can evade. The Copernican system, the Keplerian laws, and the Newtonian mechanics have subsequently proved themselves—in the interpreting of phenomena, in observational findings, and in technology—over such a range and with such extreme accuracy that after Newton's *Principia* it was no longer possible to doubt that they were correct. Yet even here there was still an idealization involved, such as Plato had held necessary and Aristotle had disapproved.

This only came out in full clarity some fifty years ago when it was realized from the findings in atomic physics that the Newtonian scheme of concepts was no longer adequate to cope with the mechanical phenomena in the interior of the atom. Since Planck's discovery of the quantum of action, in 1900, a state of confusion had arisen in physics. The old rules, whereby nature had been successfully described for more than two centuries, would no longer fit the new findings. But even these findings were themselves inherently contradictory. A hypothesis that proved itself in one experiment failed in another. The beauty and completeness of the old physics seemed destroyed, without anyone having been able, from the often disparate experiments, to gain a real insight into new and different sorts of connection. I don't know if it is fitting to compare the state of physics in those twenty-five years after Planck's discovery (which I, too, encountered as a young student) to the circumstances of contemporary modern art. But I have to confess that this comparison repeatedly comes to my mind. The helplessness when faced with the question of what to do about the bewildering phenomena, the lamenting over lost connections, which still continue to look so very convincing—all these discontents have shaped the face of both disciplines and both periods, different as they are, in a similar manner. We are obviously concerned here with a necessary intervening stage, which cannot be bypassed and which is preparing for developments to come. For, as Pauli told us, all understanding is a protracted affair, inaugurated by processes in the unconscious long before the content of consciousness can be rationally formulated.

At that moment, however, when the true Ideas rise up, there occurs in the soul of him who sees them an altogether indescribable process of the highest intensity. It is the amazed awe that Plato speaks of in the *Phaedrus,* with which the soul remembers, as it were, something it had unconsciously possessed all along. Kepler says: *"Geometria est archetypus pulchritudinis mundi"*; or, if we may translate in more general terms: "Mathematics is the archetype of the beauty of the world." In atomic physics this process took place not quite fifty years ago and has again restored exact science, under entirely new presuppositions, to that state of harmonious completeness which for a quarter of a century it had lost. I see no reason why the same thing should not also happen one day in art. But it must be added, by way of warning, that such a thing cannot be made to happen—it has to occur on its own.

I have set this aspect of exact science before you because in it the affinity with the fine arts becomes most plainly visible and because here one may counter the misapprehension that natural science and technology are concerned solely with precise observation and rational, discursive thought. To be sure, this rational thinking and careful measurement belong to the scientist's work, just as the hammer and chisel belong to the work of the sculptor. But in both cases they are merely the tools and not the content of the work.

Perhaps at the very end I may remind you once more of the second definition of the concept of beauty, which stems from Plotinus and in which no more is heard of the parts and the whole: "Beauty is the translucence, through the material phenomenon, of the eternal splendor of the 'one.'" There are important periods of art in which this definition is more appropriate than the first, and to such periods we often look longingly back. But in our own time it is hard to speak of beauty from this aspect, and perhaps it is a good rule to adhere to the custom of the age one has to live in, and to keep silent about that which it is difficult to say. In actual fact, the two definitions are not so very widely removed from one another. So let us be content with the first and more sober definition of beauty, which certainly is also realized in natural science, and let us declare that in exact science, no less than in the arts, it is the most important source of illumination and clarity.

6

If Science Is Conscious of Its Limits . . .

I T MAY BE RELEVANT to discuss the *concept of scientific truth* more generally and to enquire what are the criteria which allow us to call scientific knowledge consistent and final. Let us begin with a purely external criterion. As long as any sphere of mental life advances continuously and without any inner break, those who work in this sphere will always pose detailed questions on what we may call problems of technique, whose solution is not a purpose in itself but whose value stems from the part they play in the larger framework which alone is important. Perhaps this is the reason why sculptors in the Middle Ages tried to give the best possible descriptions of folds in dresses, the solution of this particular problem being important since even the folds in the cloaks of the saints were a part of the great religious framework which was all the artist was really concerned about. Similarly, we find that modern science continues to pose specific problems and that work on them is the condition for an understanding of the larger framework. Even in the developments of the last fifty years particular problems constantly arose by themselves. They did not have to be looked for and the aim was always that same great framework of natural laws. In this respect, and speaking purely from an external point of view, we can see no reason for any break in the continuity of the exact sciences.

With respect to the finality of the results, however, we must remind the reader that in the realm of the exact sciences there have always been final solutions for certain limited domains of experience. Thus, for instance, the questions posed by Newton's concept of mechanics found an answer valid for all time in Newton's law and in its mathematical consequences. Newtonian mechanics cannot be improved in any way, for inasmuch as we can describe a particular phenomenon with the concepts of Newtonian physics—namely, position, velocity, acceleration, mass, force, etc.—Newton's laws hold quite rigorously and nothing in this will be changed for the next hundred thousand years. More precisely, I ought perhaps to say: Newton's laws are valid to that degree of accuracy to which the phenomena concerned can be described by these concepts. The fact that this accuracy has limits was, of course, well known even to classical physicists, none of whom ever claimed he could measure to any desired degree of accuracy. The fact, however, that the accuracy of measurements is limited in principle, i.e., by uncertainty relations, is something quite new, something we first encountered in the atomic field. But, for the moment, we need not enter into this subject. For the purposes of our discussion, it is enough to assert that, inasmuch as it is possible to make accurate measurements of this kind at all, Newtonian mechanics is fully valid now and will remain so in the future.

With the reservations mentioned, it is therefore possible to say that Newtonian mechanics is a completed theory. Such a closed-off theory is characterized by a system of definitions and axioms that establishes the fundamental concepts and their interrelations, and also by the requirement that there is a wide realm of experiences, of observable phenomena, that can be described with high accuracy by means of this system. The theory is then the idealization, valid for the time, of this realm of experience.

But there are other realms of experience, and hence other closed-off theories as well. In the nineteenth century, the theory of heat, in particular, took on final form, in this sense, as a statistical statement about systems with very many degrees of freedom. The fundamental axioms of this theory define and connect such concepts as temperature, entropy, and energy, of which the first two, temperature and entropy, make no appearance whatever in Newtonian mechanics, while the last, energy, plays an important role in every field of experience and not merely in mechanics. Since the work of Willard Gibbs, the statistical theory of heat can likewise be reckoned a final and closed-off theory, nor can we doubt that its laws apply everywhere, at all times, with the highest accuracy—although naturally only to those phenomena which can be dealt with by means of such concepts as temperature, entropy, and energy. This theory, too, is an idealization; we know that there are many conditions, e.g., of matter in the gaseous state, where one cannot speak of temperature and so cannot apply the laws of this heat theory either.

From what has been said, it will already be clear that in physics, at all events, there do exist closed-off theories, which can be regarded as idealizations for restricted fields of experience and which claim to be valid for all time. But there can obviously be no talk here, as yet, of any closing off of physics as a whole.

In the last two hundred years, quite new fields of experience have been opened up by experiment. Since the foundational inquiries of Luigi Galvani and Alessandro Volta, the phenomena of electromagnetism have been studied with ever greater exactness, their relationships to chemistry being demonstrated by Faraday and those to optics by Heinrich Hertz. The fundamental facts of atomic physics were first disclosed by findings in chemistry and then explored in every detail by experiments in electrolysis, in discharge processes in gases, and later, in radioactivity. For an understanding of this gigantic new territory, the closed-off theories of an earlier day were inadequate. And so new and more comprehensive theories were framed, which can be regarded as idealizations of these new regions of experience. The theory of relativity emerged from the electrodynamics of moving bodies and has led to new insights into the structure of space and time. The quantum theory gives an account of the mechanical processes in the interior of the atom, but it also incorporates Newtonian mechanics, as the limiting case in which we are able to objectify the events completely and can neglect the interaction between the object under investigation and the observer himself.

Relativity theory, no less than quantum mechanics, can also be viewed as a closed-off theory, a very comprehensive idealization of exceedingly large tracts of experience, of whose laws we can take it that they are valid everywhere and at all times—but again only for those areas of experience which can be apprehended by means of these concepts.

Accordingly, in the exact sciences the word "final" obviously means that there are always self-contained, mathematically representable systems of concepts and laws applicable to certain realms of experience, in which realms they are always valid for the entire cosmos and cannot be changed or improved. Obviously, however, we cannot expect these concepts and laws to be suitable for the subsequent description of new realms of experience. It is only in this limited sense that quantum-theoretical concepts and laws can be considered as final, and only in this limited sense can it ever happen that scientific knowledge is finally formulated in mathematical or, for that matter, in any other language.

Similarly, many legal philosophies assume that while Law always exists, each new case generally involves a new discovery of law, that the written law can be relevant only to limited realms of life, and that it cannot be binding forever. The exact sciences also start from the assumption that in the end, it will always be possible to understand nature, even in every new field of experience, but that we may make no *a priori* assumptions about the meaning of the word "understand." In the sciences, we find that the mathematical formulations of previous epochs are "final" but by no means universal. It is because of this that it is impossible to base acts of faith, supposed to be binding for our behavior in life, on our scientific understanding alone since formulations of scientific knowledge apply only to a limited range of experience. Many modern creeds which claim that they are, in fact, not dealing with questions of faith, but are based on scientific knowledge, contain inner contradictions and rest on self-deception.

As we become clearer about this limitation, the limitation itself may be considered to be the first foothold from which we may reorientate ourselves.

Perhaps this analogy will help us in gaining a new hope that although these limitations affect us in some ways, they do not limit life itself. The space in which man develops as a spiritual being has more dimensions than the single one which it has occupied during the last centuries. This would imply that over longer periods of time, a conscious acceptance of this limitation might well lead to some equilibrium, where man's knowledge and creative forces will once again find themselves ranged spontaneously about their common center.

By way of conclusion, I shall quote the introduction to the *Principles of Mechanics* (1876) by Heinrich Hertz (1857–1894), for here it emerges clearly how physics began to remember once more that a natural science is one *whose propositions on limited domains of nature can have only a correspondingly limited validity; that science is not a philosophy developing a worldview of nature as a whole or about the essence of things.* Hertz points out that propositions in physics have neither the task nor the capacity of revealing the inherent essence of natural phenomena. He concludes that physical determinations are only pictures, on whose correspondence with natural objects we can make but the single assertion, *viz.*, whether or not the *logically* derivable consequences of our pictures correspond with the empirically observed consequences of the phenomena for which we have designed our picture. In other words, the hypothetical picture of a causal relationship with which we invest natural phenomena must prove its usefulness in practice. The criteria for assessing the suitability of a picture are that (1) it must be *admissible, i.e.,* correspond with our laws of thought; (2) it must be *correct, i.e.,* agree with experience; (3) it must be *relevant, i.e.,* contain the maximum of essential and the minimum of superfluous or empty relations of the object.

Here already we get a foretaste of the essential insight of modern physics stated with such impressive brevity by Eddington: "We have found that where science has progressed the farthest, the mind has but regained from nature that which the mind has put into nature. We have found a strange footprint on the shores of the unknown. We have devised profound theories, one after another, to account for its origin. At last, we have succeeded in reconstructing the creature that made the footprint. And Lo! it is our own."

I should like to stress the following:

1. Modern science, in its beginnings, was characterized by a conscious modesty; it made statements about strictly limited relations that *are only valid within the framework of these limitations.*
2. *This modesty was largely lost during the nineteenth century.* Physical knowledge was considered to make assertions about nature as a whole. Physics wished to turn philosopher, and the demand was voiced from many quarters that all true philosophers must be scientific.
3. Today physics is undergoing a basic change, the most characteristic trait of which is a return to its original self-limitation.
4. The philosophic content of a science is only preserved if science is conscious of its limits. Great discoveries of the properties of individual phenomena are possible only if the nature of the phenomena is not generalized *a priori.* Only by leaving open the question of the ultimate essence of a body, of matter, of energy, *etc.,* can physics reach an understanding of the individual properties of the phenomena that we designate by these concepts, an understanding which alone may lead us to real philosophical insight.

SCHROEDINGER

ERWIN SCHROEDINGER
(1887–1961)

A T ABOUT THE SAME TIME that Heisenberg et al. were developing matrix mechanics, Erwin Schroedinger independently discovered a form of "wave mechanics" that was quickly shown to be equivalent to, but in many respects simpler and more elegant than, the matrix mechanics. It was therefore "Schroedinger's wave equation" that soon became the heart of modern quantum mechanics and its most widely used mathematical tool. For this seminal work, Schroedinger was awarded the 1933 Nobel Prize in Physics.

The following sections are taken from *My View of the World* (Cambridge University Press ["C.U.P."], 1964), *Mind and Matter* (C.U.P., 1958), *Nature and the Greeks* (C.U.P., 1954), *Science and Humanism* (C.U.P., 1951), and *What Is Life?* (C.U.P., 1947). Schroedinger's mystical insight, I believe, was probably the keenest of any in this volume, and his eloquence was matched only by Eddington's. The last selection (Chapter 10), in particular, contains some of the finest and most poetic mystical statements ever penned, and stands eloquently as its own remark.

7

Why Not Talk Physics?

T HERE IS ONE COMPLAINT which I shall not escape. Not a word is said here of acausality, wave mechanics, indeterminacy relations, complementarity, an expanding universe, continuous creation, etc. Why doesn't he talk about what he knows instead of trespassing on the professional philosopher's preserves? *Ne sutor supra crepidam.* On this I can cheerfully justify myself: because I do not think that these things have as much connection as is currently supposed with a philosophical view of the world. I think that I see eye-to-eye here, on certain essential points, with Max Planck and Ernst Cassirer.

I do not think I am prejudiced against the importance that science has from the purely human point of view. But with all that, I cannot believe (and this is my first objection)—I cannot believe that [for example] the deep philosophical enquiry into the relation between subject and object and into the true meaning of the distinction between them depends on the quantitative results of physical and chemical measurements with weighing scales, spectroscopes, microscopes, telescopes, with Geiger-Müller-counters, Wilson-chambers, photographic plates, arrangements for measuring the radioactive decay, and whatnot. It is not very easy to say *why* I do not believe it. I feel a certain incongruity between the applied means and the problem to be solved. I do *not* feel quite so diffident with regard to other sciences, in particular biology, and quite especially *genetics*, and the facts about *evolution*. But we shall not talk about this here and now.

On the other hand (and this is my second objection), the mere contention that every observation depends on both the subject and the object, which are inextricably interwoven, is hardly new; it is almost as old as science itself. Though but scarce reports and quotations of the two great men from Abdera, Protagoras and Democritus, have come down to us across the twenty-four centuries that separate us from them, we can tell that they both, in their way, maintained that all our sensations, perceptions, and observations have a strong personal, subjective tinge and do not convey the nature of the thing-in-itself. (The difference between them was that Protagoras dispensed with the thing-in-itself; to him, our sensations were the only truth, while Democritus thought differently.) Since then the question has turned up whenever there was science; we might follow it through the centuries, speaking of Descartes', Leibnitz's, Kant's attitudes toward it. We shall not do this. But I must mention one point, in order not to be accused of injustice towards the quantum physicists of our days. I said their statement that in perception and observation subject and object are inextricably interwoven is hardly new. But they could make a case that something about it *is* new. I think it is true that in previous centuries, when discussing this question, one mostly had in mind two things, viz. (a) a direct physical *impression caused* by the object in the subject, and (b) the *state* of the subject that receives the impression. As against this, in the present order of ideas the direct physical, causal, influence between the two is regarded as *mutual*. It is said that there is also an unavoidable and uncontrollable impression from the side of the *subject* onto the *object*. This aspect *is* new, and, I should say, more adequate anyhow. For physical action always is *inter*-action; it always *is* mutual. What remains doubtful to me is only just this: whether it is adequate to term one of the two physically interacting systems the "subject." *For the observing mind is not a physical system, it cannot interact with any physical system.* And it might be better to reserve the term "subject" for the observing mind.

[In other words, Schroedinger acknowledges that quantum mechanics shows, if anything, an interaction between *objects*, not between subject and object. The reason he denies the latter—and the reason he seems to have so little use for the alleged impact of quantum interaction on philosophy and mysticism—is explained in the following paragraphs.—Ed. note]

But from the theory as explained before, from the unavoidable and unsurveyable interference of the measuring devices with the object under observation, lofty consequences of an epistemological nature have been drawn and brought to the fore, concerning the relation between subject and object. It is maintained that recent discoveries in physics have pushed forward to the mysterious boundary between the subject and the object. This boundary, so we are told, is not a sharp boundary at all. We are given to understand that we never observe an object without its being modified or tinged by our own activity in observing it. We are given to understand that under the impact of our refined methods of observation and of thinking about the results of our experiments that mysterious boundary between the subject and the object has broken down.

In order to criticize these contentions let me at first accept the time-hallowed distinction or discrimination between object and subject, as many thinkers both in olden times have accepted it and in recent times still accept it. Among the philosophers who accepted it—from Democritus of Abdera down to the "Old Man of Königsberg"—there were few, if any, who did not emphasize that all our sensations, perceptions, and observations have a strong, personal, subjective tinge and do not convey the nature of the "thing-in-itself," to use Kant's term. While some of these thinkers might have in mind only a more or less strong or slight distortion, Kant landed us with a complete resignation: never to know anything at all about his "thing-in-itself." Thus the idea of subjectivity in all appearance is very old and familiar. What is new in that present setting is this: that not only would the impressions we get from our environment largely depend on the nature and the contingent state of our sensorium, but, inversely, the very environment that we wish to take in is modified by us, notably by the devices we set up in order to observe it.

Maybe this is so—to some extent it certainly is. Maybe that from the newly discovered laws of quantum physics this modification cannot be reduced below certain well-ascertained limits. *Still I would not like to call this a direct influence of the subject on the object.* For the subject, if anything, is the thing that senses and thinks. Sensations and thoughts do not belong to the "world of energy." They cannot produce any change in this world of energy as we know from Spinoza and Sir Charles Sherrington.

The same elements compose my mind and the world. This situation is the same for every mind and its world, in spite of the unfathomable abundance of "cross-references" between them. The world is given to me only once, not one existing and one perceived. Subject and object are only one. The barrier between them cannot be said to have broken down as a result of recent experience in the physical sciences, for this barrier does not exist.

COULD PHYSICAL INDETERMINACY GIVE FREE WILL A CHANCE?

Could perhaps the declared *indeterminacy* allow *free will* to step into the gap in the way that *free will determines* those events which the Law of Nature leaves undetermined? This hope is, at first sight, obvious and understandable.

In this crude form the attempt was made, and the idea, to a certain extent, worked out by the German physicist Pascual Jordan. I believe it to be both physically and morally an impossible solution. As regards the first: according to our present view, the quantum laws, though they leave the single event undetermined, predict a quite definite *statistics* of events when the same situation occurs again and again. If these statistics are interfered with by any agent, this agent violates the laws of quantum mechanics just as objectionably as if it interfered—in pre-quantum physics—with a strictly causal mechanical law. Now we know that *there are not statistics* in the reaction of the same person to precisely the same moral situation—the rule is that the same individual in the same situation acts again precisely in the same manner. (Mind you, in *precisely* the same situation; this does not mean that a criminal or addict cannot be converted or healed by persuasion and example or whatnot—by strong external influence. But this, of course, means that the situation is changed.) The inference is that Jordan's assumption—the direct stepping in of free will to fill the gap of indeterminacy—does amount to an interference with the laws of nature, even in their form accepted in quantum theory. But at *that* price, of course, we can have everything. This is not a solution of the dilemma.

The moral objection was strongly emphasized by the German philosopher Ernst Cassirer (who died in 1945 in New York as an exile from Nazi Germany). Cassirer's extended criticism of Jordan's ideas is based on a thorough familiarity with the situation in physics. I shall try to summarize it briefly; I would say it amounts to this. Free will in man includes as its most relevant part man's ethical behavior. Supposing the physical events in space and time actually are to a large extent not strictly determined but subject to pure chance, as most physicists in our time believe, then this haphazard side of the goings-on in the material world is certainly (says Cassirer) *the very last to be invoked as the physical correlate of man's ethical behavior.* For this is anything but haphazard; it is intensely determined by motives ranging from the lowest to the most sublime sort, from greed and spite to genuine love of the fellow creature or sincere religious devotion. Cassirer's lucid discussion makes one feel so strongly the absurdity of basing free will, including ethics, on physical haphazard that the previous difficulty, the antagonism between free will and determinism, dwindles and almost vanishes under the mighty blows Cassirer deals to the opposite view. "Even the reduced extent of predictability" (Cassirer adds) "still granted by Quantum Mechanics would amply suffice to destroy ethical freedom, if the concept and true meaning of the latter were irreconcilable with predictability." Indeed, one begins to wonder whether the supposed paradox is really so shocking, and whether physical determinism is not perhaps quite a suitable correlate to the mental phenomenon of will, which is not always easy to predict "from outside," but usually extremely determined "from inside." To my mind, this is the most valuable outcome of the whole controversy: the scale is turned in favour of a possible reconciliation of free will with physical determinism, when we realize how inadequate a basis physical haphazard provides for ethics.

The net result is that quantum physics has nothing to do with the free will problem. If there is such a problem, it is not furthered a whit by the latest development in physics. To quote Ernst Cassirer again: "Thus it is clear . . . that a possible change in the physical concept of causality can have no immediate bearing on ethics."

Science Cannot Touch It

The scientific picture of the real world around me is very deficient. It gives a lot of factual information, puts all our experience in a magnificently consistent order, but it is ghastly silent about all and sundry that is really near to our heart, that really matters to us. It cannot tell us a word about red and blue, bitter and sweet, physical pain and physical delight; it knows nothing of beautiful and ugly, good or bad, God and eternity. Science sometimes pretends to answer questions in these domains, but the answers are very often so silly that we are not inclined to take them seriously.

So, in brief, we do not belong to this material world that science constructs for us. We are not in it; we are outside. We are only spectators. The reason why we believe that we are in it, that we belong to the picture, is that our bodies are in the picture. Our bodies belong to it. Not only my own body, but those of my friends, also of my dog and cat and horse, and of all the other people and animals. And this is my only means of communicating with them.

Moreover, my body is implied in quite a few of the more interesting changes—movements, etc.—that go on in this material world, and is implied in such a way that I feel myself partly the author of these goings-on. But then comes the impasse, this very embarrassing discovery of science, that I am not needed as an author. Within the scientific world-picture all these happenings take care of themselves—they are amply accounted for by direct energetic interplay. Even the human body's movements "are its own" as Sherrington put it. The scientific world-picture vouchsafes a very complete understanding of all that happens—it makes it just a little too understandable. It allows you to imagine the total display as that of a mechanical clockwork which, for all that science knows, could go on just the same as it does, without there being consciousness, will, endeavor, pain and delight and responsibility connected with it—though they actually are. And the reason for this disconcerting situation is just this: that, for the purpose of constructing the picture of the external world, we have used the greatly simplifying device of cutting our own personality out, removing it; hence it is gone, it has evaporated, it is ostensibly not needed.

In particular, and most importantly, this is the reason why the scientific worldview contains of itself no ethical values, no aesthetical values, not a word about our own ultimate scope or destination, and no God, if you please. Whence came I, whither go I?

Science cannot tell us a word about why music delights us, of why and how an old song can move us to tears.

Science, we believe, can, in principle, describe in full detail all that happens in the latter case in our sensorium and "motorium" from the moment the waves of compression and dilation reach our ear to the moment when certain glands secrete a salty fluid that emerges from our eyes. But of the feelings of delight and sorrow that accompany the process science is completely ignorant—and therefore, reticent.

Science is reticent too when it is a question of the great Unity—the One of Parmenides—of which we all somehow form part, to which we belong. The most popular name for it in our time is God—with a capital "G." Science is, very usually, branded as being atheistic. After what we said, this is not astonishing. If its world-picture does not even contain blue, yellow, bitter, sweet—beauty, delight, and sorrow—, if personality is cut out of it by agreement, how should it contain the most sublime idea that presents itself to human mind?

The world is big and great and beautiful. My scientific knowledge of the events in it comprises hundreds of millions of years. Yet in another way it is ostensibly contained in a poor seventy or eighty or ninety years granted to me—a tiny spot in immeasurable time, nay even in the finite millions and milliards of years that I have learnt to measure and to assess. Whence come I and whither go I? That is the great unfathomable question, the same for every one of us. Science has no answer to it.

8

The Oneness of Mind

THE REASON WHY our sentient, percipient, and thinking ego is met nowhere within our scientific world picture can easily be indicated in seven words: because it is itself that world picture. It is identical with the whole and therefore cannot be contained in it as a part of it. But, of course, here we knock against the arithmetical paradox; there appears to be a great multitude of these conscious egos, the world, however, is only one. This comes from the fashion in which the world-concept produces itself. The several domains of "private" consciousnesses partly overlap. The region common to all where they all overlap is the construct of the "real world around us." With all that an uncomfortable feeling remains, prompting such questions as: is my world really the same as yours? Is there *one* real world to be distinguished from its pictures introjected by way of perception into every one of us? And if so, are these pictures like unto the real world or is the latter, the world "in itself," perhaps very different from the one we perceive?

Such questions are ingenious, but, in my opinion, very apt to confuse the issue. They have no adequate answers. They all are, or lead to, antinomies springing from the one source, which I called the arithmetical paradox; the *many* conscious egos from whose mental experiences the *one* world is concocted. The solution of this paradox of numbers would do away with all the questions of the aforesaid kind and reveal them, I dare say, as sham-questions.

There are two ways out of the number paradox, both appearing rather lunatic from the point of view of present scientific thought (based on ancient Greek thought and thus thoroughly "Western"). One way out is the multiplication of the world in Leibniz's fearful doctrine of

86

monads: every monad to be a world by itself, no communication between them; the monad "has no windows," it is "incommunicado." That, nonetheless, they all agree with each other is called "pre-established harmony." I think there are few to whom this suggestion appeals, nay who would consider it as a mitigation at all of the numerical antinomy.

There is obviously only one alternative, namely the unification of minds or consciousnesses. Their multiplicity is only apparent, in truth, there is only one mind. This is the doctrine of the Upanishads. And not only of the Upanishads. The mystically experienced union with God regularly entails this attitude unless it is opposed by strong existing prejudices; this means that it is less easily accepted in the West than in the East. Let me quote, as an example outside the Upanishads, an Islamic-Persian mystic of the thirteenth century, Aziz Nasafi. I am taking it from a paper by Fritz Meyer and translating from his German translation:

> On the death of any living creature the spirit returns to the spiritual world, the body to the bodily world. In this however only the bodies are subject to change. The spiritual world is one single spirit who stands like unto a light behind the bodily world and who, when any single creature comes into being, shines through it as through a window. According to the kind and size of the window less or more light enters the world. The light itself however remains unchanged.

Ten years ago, Aldous Huxley published a precious volume which he called *The Perennial Philosophy* and which is an anthology from the mystics of the most various periods and the most various peoples. Open it where you will and you find many beautiful utterances of a similar kind. You are struck by the miraculous agreement between humans of different race, different religion, knowing nothing about each other's existence, separated by centuries and millenia, and by the greatest distances that there are on our globe.

One thing can be claimed in favour of the mystical teaching of the "identity" of all minds with each other and with the supreme mind—as against the fearful monadology of Leibniz. The doctrine of identity can claim that it is clinched by the empirical fact that consciousness is never experienced in the plural, only in the singular. Not only has none of us ever experienced more than one consciousness, but there is also no trace of circumstantial evidence of this ever happening anywhere in the world.

If I say that there cannot be more than one consciousness in the same mind, this seems a blunt tautology—we are quite unable to imagine the contrary.

Yet there are cases or situations where we would expect and nearly require this unimaginable thing to happen, if it can happen at all. This is the point that I should like to discuss now in some detail, and to clinch it by quotations from Sir Charles Sherrington, who was at the same time (rare event!) a man of highest genius and a sober scientist. I will give you the main conclusion in Sherrington's own words:

> It is not spatial conjunction of cerebral mechanism which combines the two reports. . . . It is much as though the right-and left-eye images were seen each by one of two observers and the minds of the two observers were combined to a single mind. It is as though the right-eye and left-eye perceptions are elaborated singly and then psychically combined to one. . . . It is as if each eye had a separate sensorium of considerable dignity proper to itself, in which mental processes based on that eye were developed up to even full perceptual levels. Such would amount physiologically to a visual sub-brain. There would be two such sub-brains, one for the right eye and one for the left eye. Contemporaneity of action rather than structural union seems to provide their mental collaboration.

This is followed by very general considerations, of which I shall again pick out only the most characteristic passages:

> Are there thus quasi-independent sub-brains based on the several modalities of sense? In the roof-brain the old "five" senses instead of being merged inextricably in one another and further submerged under mechanism of higher order are still plain to find, each demarcated in its separate sphere. How far is the mind a collection of quasi-independent perceptual minds integrated psychically in large measure by temporal concurrence of experience? . . . When it is a question of "mind" the nervous system does not integrate itself by centralization upon a pontifical cell. Rather it elaborates a million-fold democracy whose each unit is a cell . . . the concrete life compounded of sublives reveals, although integrated, its additive nature and declares itself an affair of minute foci of life acting together. . . . When however we

turn to the mind there is nothing of all this. The single nerve-cell is never a miniature brain. The cellular constitution of the body need not be for any hint of it from "mind". . . . A single pontifical brain-cell could not assure to the mental reaction a character more unified, and non-atomic than does the roof-brain's multitudinous sheet of cells. Matter and energy seem granular in structure, and so does "life," but not so mind.

I have quoted you the passages which have most impressed me. Sherrington, with his superior knowledge of what is actually going on in a living body, is seen struggling with a paradox which, in his candidness and absolute intellectual sincerity, he does not try to hide away or explain away (as many others would have done, nay have done), but he almost brutally exposes it, knowing very well that this is the only way of driving any problem in science or philosophy nearer towards its solution, while by plastering it over with "nice" phrases you prevent progress and make the antinomy perennial (not forever, but until someone notices your fraud). Sherrington's paradox too is an arithmetical paradox, a paradox of numbers, and it has, so I believe, very much to do with the one to which I had given this name earlier in this chapter, though it is by no means identical with it. The previous one was briefly, the *one* world crystallizing out of the many minds. Sherrington's is the *one* mind, based ostensibly on the many cell-lives or, in another way, on the manifold sub-brains, each of which seems to have such a considerable dignity proper to itself that we feel impelled to associate a sub-mind with it. Yet we know that a sub-mind is an atrocious monstrosity, just as is a plural-mind—neither having any counterpart in anybody's experience, neither being in any way imaginable. Mind is, by its very nature, a *singulare tantum*. I should say: the overall number of minds is just one. I venture to call it indestructible since it has a peculiar timetable, namely mind is always *now*. There is really no before and after for mind. There is only a now that includes memories and expectations. But I grant that our language is not quite adequate to express this, and I also grant, should anyone wish to state it, that I am now talking religion, not science. . . .

Sherrington says: "Man's mind is a recent product of our planet's side."

I agree, naturally. If the first word (man's) were left out, I would not. It would seem queer, not to say ridiculous, to think that the contemplating, conscious mind that alone reflects the becoming of the world should have made its appearance only at some time in the course of this "becoming," should have appeared contingently, associated with a very special biological contraption which, in itself, quite obviously discharges the task of facilitating certain forms of life in maintaining themselves, thus favoring their preservation and propagation: forms of life that were latecomers and have been preceded by many others that maintained themselves without that particular contraption (a brain). Only a small fraction of them (if you count by species) have embarked on "getting themselves a brain." And before that happened, should it all have been a performance to empty stalls? Nay, may we call a world that nobody contemplates even that? When an archeologist reconstructs a city or a culture long bygone, he is interested in human life in the past, in actions, sensations, thoughts, feelings, in joy and sorrow of humans, displayed there and then. But a world, existing for many millions of years without any mind being aware of it, contemplating it, is it anything at all? Has it existed? For do not let us forget: to say, as we did, that the becoming of the world is reflected in a conscious mind is but a cliché, a phrase, a metaphor that has become familiar to us. The world is given but once. Nothing is reflected. The original and the mirror-image are identical. The world extended in space and time is but our representation *(Vorstellung)*. Experience does not give us the slightest clue of its being anything besides that—as Berkeley was well aware.

Sometimes a painter introduces into his large picture, or a poet into his long poem, an unpretending subordinate character who is himself. Thus the poet of the *Odyssey* has, I suppose, meant himself by the blind bard who in the hall of the Phaeacians sings about the battles of Troy and moves the battered hero to tears. In the same way we meet in the song of the Nibelungs, when they traverse the Austrian lands with a poet who is suspected to be the author of the whole epic. In Dürer's *All-Saints* picture, two circles of believers are gathered in prayer around the Trinity high up in the skies: a circle of the blessed above and a circle of humans on the earth. Among the latter are kings and emperors and popes, but also, if I am not mistaken, the portrait of the artist himself, as a humble side figure that might as well be missing.

To me, this seems to be the best simile of the bewildering double role of mind. On the one hand, mind is the artist who has produced the whole; in the accomplished work, however, it is but an insignificant accessory that might be absent without detracting from the total effect.

Speaking without metaphor, we have to declare that we are here faced with one of these typical antinomies caused by the fact that we have not yet succeeded in elaborating a fairly understandable outlook on the world without retiring our own mind, the producer of the world picture, from it, so that mind has no place in it. The attempt to press it into it, after all, necessarily produces some absurdities.

Earlier, I have commented on the fact that, for this same reason, the physical world picture lacks all the sensual qualities that go to make up the Subject of Cognizance. The model is colorless and soundless and unpalpable. In the same way and for the same reason, the world of science lacks, or is deprived of, everything that has a meaning only in relation to the consciously contemplating, perceiving, and feeling subject. I mean, in the first place, the ethical and aesthetical values, any values of any kind, everything related to the meaning and scope of the whole display. *All this is not only absent but it cannot, from the purely scientific point of view, be inserted organically.* If one tries to put it in or on, as a child puts color on his uncolored painting copies, it will not fit. For anything that is made to enter this world-model willy-nilly takes the form of scientific assertion of facts, and as such, it becomes wrong.

Most painful is the absolute silence of all our scientific investigations toward our questions concerning the meaning and scope of the whole display. The more attentively we watch it, the more aimless and foolish it appears to be. The show that is going on obviously acquires a meaning only with regard to the mind that contemplates it. But what science tells us about this relationship is patently absurd: as if mind had only been produced by that very display that is now watching and would pass away with it when the sun finally cools down and the earth has been turned into a desert of ice and snow.

Let me briefly mention the notorious atheism of science which comes, of course, under the same heading. Science has to suffer this reproach again and again, but unjustly so. No personal god can form part of a world-model that has only become accessible at the cost of removing everything personal from it. We know, when God is experienced, this is an event as real as an immediate sense perception or as one's own personality. Like them, he must be missing in the space-time picture. I do not find God anywhere in space and time—that is what the honest naturalist tells you. For this, he incurs blame from him in whose catechism is written: God is spirit.

9

The I That Is God

A s a reward for the serious trouble I have taken to expound the purely scientific aspect of our problem *sine ira et studio,* I beg leave to add my own, necessarily subjective, view of its philosophical implications.

According to the evidence put forward in the preceding pages, the space-time events in the body of a living being which correspond to the activity of its mind, to its self-conscious or any other actions, are (considering also their complex structure and the accepted statistical explanation of physico-chemistry) if not strictly deterministic at any rate statistico-deterministic. To the physicist, I wish to emphasize that in my opinion, and contrary to the opinion upheld in some quarters, *quantum indeterminacy* plays no biologically relevant role in them, except perhaps by enhancing their purely accidental character in such events as meiosis, natural and X-ray-induced mutation and so on—and this is, in any case, obvious and well recognized.

For the sake of argument, let me regard this as a fact, as I believe every unbiased biologist would, if there were not the well-known, unpleasant feeling about "declaring oneself to be a pure mechanism." For it is deemed to contradict Free Will as warranted by direct introspection.

But immediate experiences in themselves, however various and disparate they be, are logically incapable of contradicting each other. So let us see whether we cannot draw the correct, non-contradictory conclusion from the following two premises:

(i) My body functions as a pure mechanism according to the Laws of Nature.

(ii) Yet I know, by incontrovertible direct experience, that I am directing its motions, of which I foresee the effects, that may be fateful and all-important, in which case I feel and take full responsibility for them.

The only possible inference from these two facts is, I think, that I—I in the widest meaning of the word, that is to say, every conscious mind that has ever said or felt "I"—am the person, if any, who controls the "motion of the atoms" according to the Laws of Nature.

Within a cultural milieu *(Kulturkreis)* where certain conceptions (which once had or still have a wider meaning amongst other peoples) have been limited and specialized, it is daring to give to this conclusion the simple wording that it requires. In Christian terminology to say: "Hence I am God Almighty" sounds both blasphemous and lunatic. But please disregard these connotations for the moment and consider whether the above inference is not the closest a biologist can get to proving God and immortality at one stroke.

In itself, the insight is not new. The earliest records, to my knowledge, date back some 2500 years or more. From the early great Upanishads the recognition ATMAN = BRAHMAN (the personal self equals the omnipresent, all-comprehending eternal self) was in Indian thought considered, far from being blasphemous, to represent the quintessence of deepest insight into the happenings of the world. The striving of all the scholars of Vedanta was, after having learnt to pronounce with their lips, really to assimilate in their minds this grandest of all thoughts.

Again, the mystics of many centuries, independently, yet in perfect harmony with each other (somewhat like the particles in an ideal gas) have described, each of them, the unique experience of his or her life in terms that can be condensed in the phrase: DEUS FACTUS SUM (I have become God).

To Western ideology, the thought has remained a stranger, in spite of Schopenhauer and others who stood for it and in spite of those true lovers who, as they look into each other's eyes, become aware that their thought and their joy are *numerically* one, not merely similar or identical—but they, as a rule, are emotionally too busy to indulge in clear thinking, in which respect they very much resemble the mystic.

Allow me a few further comments. Consciousness is never experienced in the plural, only in the singular. Even in the pathological cases of split consciousness or double personality the two persons alternate, they are never manifest simultaneously. In a dream we do perform several characters at the same time, but not indiscriminately: we *are* one of them; in him, we act and speak directly while we often eagerly await the answer or response of another person, unaware of the fact that it is we who control his movements and his speech just as much as our own.

How does the idea of plurality (so emphatically opposed by the Upanishad writers) arise at all? Consciousness finds itself intimately connected with, and dependent on, the physical state of a limited region of matter, the body. (Consider the changes of mind during the development of the body, as puberty, aging, dotage, etc., or consider the effects of fever, intoxication, narcosis, lesion of the brain, and so on.) Now, there is a great plurality of similar bodies. Hence the pluralization of consciousnesses or minds seems a very suggestive hypothesis. Probably all simple ingenuous people, as well as the great majority of western philosophers, have accepted it.

It leads almost immediately to the invention of souls, as many as there are bodies, and to the question whether they are mortal as the body is or whether they are immortal and capable of existing by themselves. The former alternative is distasteful, while the latter frankly forgets, ignores, or disowns the facts upon which the plurality hypothesis rests. Much sillier questions have been asked: Do animals also have souls? It has even been questioned whether women, or only men, have souls.

Such consequences, even if only tentative, must make us suspicious of the plurality hypothesis, which is common to all official Western creeds. Are we not inclining to much greater nonsense if in discarding their gross superstitions, we retain their naïve idea of plurality of souls, but "remedy" it by declaring the souls to be perishable, to be annihilated with the respective bodies?

The only possible alternative is simply to keep the immediate experience that consciousness is a singular of which the plural is unknown; that there *is* only one thing and that, what seems to be a plurality, is merely a series of different aspects of this one thing, produced by a deception (the Indian MAYA); the same illusion is produced in a gallery of mirrors, and in the same way Gaurisankar and Mt. Everest turned out to be the same peak, seen from different valleys.

There are, of course, elaborate ghost stories fixed in our minds to hamper our acceptance of such simple recognition. E.g., it has been said that there is a tree there outside my window, but I do not really see the tree. By some cunning device of which only the initial, relatively simple steps are explored, the real tree throws an image of itself into my consciousness, and that is what I perceive. If you stand by my side and look at the same tree, the latter manages to throw an image into your soul as well. I see my tree and you see yours (remarkably like mine), and what the tree in itself is we do not know. For this extravagance, Kant is responsible. In the order of ideas, which regards consciousness as a *singulare tantum,* it is conveniently replaced by the statement that there is obviously only *one* tree and all the image business is a ghost story.

Yet each of us has the undisputable impression that the sum total of his own experience and memory forms a unit, quite distinct from that of any other person. He refers to it as "I." *What is this "I?"*

If you analyze it closely, you will, I think, find that it is just a little bit more than a collection of single data (experiences and memories), namely, the canvas *upon which* they are collected. And you will, on close introspection, find that what you really mean by "I," is that ground-stuff upon which they are collected. You may come to a distant country, lose sight of all your friends, may all but forget them; you acquire new friends, you share life with them as intensely as you ever did with your old ones. Less and less important will become the fact that, while living your new life, you still recollect the old one. "The youth that was I," you may come to speak of him in the third person; indeed, the protagonist of the novel you are reading is probably nearer to your heart, certainly more intensely alive and better known to you. Yet there has been no intermediate break, no death. And even if a skilled hypnotist succeeded in blotting out entirely all your earlier reminiscences, you would not find that he had killed *you.* In no case is there a loss of personal existence to deplore.

Nor will there ever be.

10

The Mystic Vision

F OR PHILOSOPHY, the real difficulty lies in the spatial and temporal multiplicity of observing and thinking individuals. If all events took place in *one* consciousness, the whole situation would be extremely simple. There would then be something given, a simple datum, and this, however otherwise constituted, could scarcely present us with a difficulty of such magnitude as the one we do, in fact, have on our hands.

I do not think that this difficulty can be logically resolved, by consistent thought, within our intellects. But it is quite easy to express the solution in words, thus: the plurality that we perceive is only *an appearance; it is not real.* Vedantic philosophy, in which this is a fundamental dogma, has sought to clarify it by a number of analogies, one of the most attractive being the many-faceted crystal which, while showing hundreds of little pictures of what is in reality a single existent object, does not really multiply that object. We intellectuals of today are not accustomed to admit a pictorial analogy as a philosophical insight; we insist on logical deduction. But, as against this, it may perhaps be possible for logical thinking to disclose at least this much: that to grasp the basis of phenomena through logical thought may, in all probability, be impossible since logical thought is itself a part of phenomena and wholly involved in them; we may ask ourselves whether, in that case, we are obliged to deny ourselves the use of an allegoric picture of the situation, merely on the grounds that its fitness cannot be strictly proved. In a considerable number of cases, logical thinking brings us up to a certain point and then leaves us in the lurch. Faced with an area not directly accessible to these lines of thought, but one into which they seem to lead, we may manage to fill it in in such a way that the lines do not simply peter out, but converge on some central point in that area; this may amount to an extremely valuable rounding-out of our picture of the world, and its worth is not to be judged by those standards of rigorous, unequivocal inescapability from which we started out. There are hundreds of cases in which science uses this procedure, and it has long been recognized as justified.

Later on, we shall try to adduce some support for the basic Vedantic vision, chiefly by pointing out particular lies in modern thought which converge upon it. Let us first be permitted to sketch a concrete picture of an *experience* which may lead toward it. In what follows, the particular situation described at the beginning could be replaced, equally fittingly, by any other; it is merely meant as a reminder that this is something that needs to be *experienced,* not simply given a notional acknowledgement.

Suppose you are sitting on a bench beside a path in high mountain country. There are grassy slopes all around, with rocks thrusting through them; on the opposite slope of the valley there is a stretch of scree with a low growth of alder bushes. Woods climb steeply on both sides of the valley, up to the line of treeless pasture; facing you, soaring up from the depths of the valley, is the mighty, glacier-tipped peak, its smooth snowfields and hard-edged rock faces touched at this moment with soft rose colour by the last rays of the departing sun, all marvellously sharp against the clear, pale, transparent blue of the sky.

According to our usual way of looking at it, everything that you are seeing has, apart from small changes, been there for thousands of years before you. After a while—not long—you will no longer exist, and the woods and rocks and sky will continue, unchanged, for thousands of years after you.

What is it that has called you so suddenly out of nothingness to enjoy for a brief while a spectacle which remains quite indifferent to you? The conditions for your existence are almost as old as the rocks. For thousands of years men have striven and suffered and begotten and women have brought forth in pain. A hundred years ago, perhaps, another man sat on this spot; like you, he gazed with awe and yearning in his heart at the dying light on the glaciers. Like you, he was begotten of man and born of woman. He felt pain and brief joy as you do. *Was* he someone else? Was it not you yourself? What is this Self of yours? What was the necessary condition for making the thing conceived this time into *you,* just *you,* and not someone else? What clearly intelligible *scientific* meaning can this "someone else" really have? If she who is now your mother had cohabited with someone else and had a son by him, and your father had done likewise, would *you* have come to be? Or were you living in them, and in your father's father, thousands of years ago? And even if this is so, why are you not your brother, why is your brother not you, why are you not one of your distant cousins? What justifies you in obstinately discovering this difference—the difference between you and someone else—when objectively what is there is *the same?*

Looking and thinking in that manner you may suddenly come to see, in a flash, the profound rightness of the basic conviction in Vedanta: it is not possible that this unity of knowledge, feeling, and choice which you call *your own* should have sprung into being from nothingness at a given moment not so long ago; rather this knowledge, feeling, and choice are essentially eternal and unchangeable and numerically *one* in all men, nay in all sensitive beings. But not in *this* sense—that *you* are a part, a piece, of an eternal, infinite being, an aspect or modification of it, as in Spinoza's pantheism. For we should then have the same baffling question: which part, which aspect are *you?* what, objectively, differentiates it from the others? No, but, inconceivable as it seems to ordinary reason, you—and all other conscious beings as such—are all in all. Hence this life of yours which you are living is not merely a piece of the entire existence, but is, in a certain sense, the *whole;* only this whole is not so constituted that it can be surveyed in one single glance. This, as we know, is what the Brahmins express in that sacred, mystic formula which is yet really so simple and so clear: *Tat tvam asi,* this is you. Or, again, in such words as "I am in the east and in the west, I am below and above, *I am this whole world."*

Thus you can throw yourself flat on the ground, stretched out upon Mother Earth, with the certain conviction that you are one with her and she with you. You are as firmly established, as invulnerable, as she— indeed, a thousand times firmer and more invulnerable. As surely as she will engulf you tomorrow, so surely will she bring you forth anew to new striving and suffering. And not merely, "some day": now, today, every day she is bringing you forth, not *once,* but thousands upon thousands of times, just as every day she engulfs you a thousand times over. For eternally and always there is only *now,* one and the same now; the present is the only thing that has no end.

EINSTEIN

ALBERT EINSTEIN
(1879–1955)

ALBERT EINSTEIN is generally regarded, quite simply, as the greatest physicist ever to have lived. His contributions to physics are legion: special and general relativity theory, quantum photoelectric effect, Brownian movement theory, the immortal $E = mc^2$. He was awarded the Nobel Prize in Physics in 1921.

The following sections are taken from *Ideas and Opinions* (New York: Crown Publishers, 1954). Einstein's mysticism has been described as a cross between Spinoza and Pythagoras; there is a central order to the cosmos, an order that can be directly apprehended by the soul in mystical union. He devoutly believed that although science, religion, art, and ethics are necessarily distinct endeavors, it is wonderment in the face of "the Mystery of the Sublime" that properly motivates them all.

11

Cosmic Religious Feeling

E VERYTHING THAT THE HUMAN RACE has done and thought is concerned with the satisfaction of deeply felt needs and the assuagement of pain. One has to keep this constantly in mind if one wishes to understand spiritual movements and their development. Feeling and longing are the motive force behind all human endeavor and human creation, in however exalted a guise the latter may present themselves to us. Now what are the feelings and needs that have led men to religious thought and belief in the widest sense of the words? A little consideration will suffice to show us that the most varying emotions preside over the birth of religious thought and experience. With primitive man it is, above all, fear that evokes religious notions—fear of hunger, wild beasts, sickness, death. Since at this stage of existence understanding of causal connections is usually poorly developed, the human mind creates illusory beings more or less analogous to itself on whose wills and actions these fearful happenings depend. Thus one tries to secure the favor of these beings by carrying out actions and offering sacrifices which, according to the tradition handed down from generation to generation, propitiate them or make them well disposed toward a mortal. In this sense, I am speaking of a religion of fear. This, though not created, is in an important degree stabilized by the formation of a special priestly caste which sets itself up as a mediator between the people and the beings they fear, and erects a hegemony on this basis. In many cases, a leader or ruler or a privileged class whose position rests on other factors combines priestly functions with its secular authority in order to make the latter more secure; or the political rulers and the priestly caste make common cause in their own interests.

The social impulses are another source of the crystallization of religion. Fathers and mothers and the leaders of larger human communities are mortal and fallible. The desire for guidance, love, and support prompts men to form the social or moral conception of God. This is the God of Providence, who protects, disposes, rewards, and punishes; the God who, according to the limits of the believer's outlook, loves and cherishes the life of the tribe or of the human race, or even life itself; the comforter in sorrow and unsatisfied longing; he who preserves the souls of the dead. This is the social or moral conception of God.

The Jewish scriptures admirably illustrate the development from the religion of fear to moral religion, a development continued in the New Testament. The religions of all civilized peoples, especially the peoples of the Orient, are primarily moral religions. The development from a religion of fear to moral religion is a great step in peoples' lives. And yet, that primitive religions are based entirely on fear and the religions of civilized peoples purely on morality is a prejudice against which we must be on our guard. The truth is that all religions are a varying blend of both types, with this differentiation: that on the higher levels of social life the religion of morality predominates.

Common to all these types is the anthropomorphic character of their conception of God. In general, only individuals of exceptional endowments, and exceptionally high-minded communities, rise to any considerable extent above this level. *But there is a third stage of religious experience* which belongs to all of them, even though it is rarely found in a pure form: I shall call it cosmic religious feeling. It is very difficult to elucidate this feeling to anyone who is entirely without it, especially as there is no anthropomorphic conception of God corresponding to it.

The individual feels the futility of human desires and aims and the sublimity and marvelous order which reveal themselves both in nature and in the world of thought. Individual existence impresses him as a sort of prison and he wants to experience the universe as a single significant whole. The beginnings of cosmic religious feeling already appear at an early stage of development, e.g., in many of the Psalms of David and in some of the Prophets. Buddhism, as we have learned especially from the wonderful writings of Schopenhauer, contains a much stronger element of this.

The religious geniuses of all ages have been distinguished by this kind of religious feeling, which knows no dogma and no God conceived in man's image; so that there can be no church whose central teachings are based on it. Hence it is precisely among the heretics of every age that we find men who were filled with this highest kind of religious feeling and were, in many cases, regarded by their contemporaries as atheists, sometimes also as saints. Looked at in this light, men like Democritus, Francis of Assisi, and Spinoza are closely akin to one another.

How can cosmic religious feeling be communicated from one person to another if it can give rise to no definite notion of a God and no theology? In my view, it is the most important function of art and science to awaken this feeling and keep it alive in those who are receptive to it.

We thus arrive at a conception of the relation of science to religion very different from the usual one. When one views the matter historically, one is inclined to look upon science and religion as irreconcilable antagonists, and for a very obvious reason. The man who is thoroughly convinced of the universal operation of the law of causation cannot for a moment entertain the idea of a being who interferes in the course of events—provided, of course, that he takes the hypothesis of causality really seriously. He has no use for the religion of fear and equally little for social or moral religion. A God who rewards and punishes is inconceivable to him for the simple reason that a man's actions are determined by necessity, external and internal, so that in God's eyes he cannot be responsible, any more than an inanimate object is responsible for the motions it undergoes. Science has, therefore, been charged with undermining morality, but the charge is unjust. A man's ethical behavior should be based effectually on sympathy, education, and social ties and needs; no religious basis is necessary. Man would indeed be in a poor way if he had to be restrained by fear of punishment and hope of reward after death.

It is, therefore, easy to see why the churches have always fought science and persecuted its devotees. On the other hand, I maintain that the cosmic religious feeling is the strongest and noblest motive for scientific research. Only those who realize the immense efforts and, above all, the devotion without which pioneer work in theoretical science cannot be achieved are able to grasp the strength of the emotion out of which alone such work, remote as it is from the immediate realities of life, can issue. What a deep conviction of the rationality of the universe and what a yearning to understand, were it but a feeble reflection of the mind revealed in this world, Kepler and Newton must have had to enable them to spend years of solitary labor in disentangling the principles of celestial mechanics! Those whose acquaintance with scientific research is derived chiefly from its practical results easily develop a completely false notion of the mentality of the men who, surrounded by a skeptical world, have shown the way to kindred spirits scattered wide through the world and the centuries. Only one who has devoted his life to similar ends can have a vivid realization of what has inspired these men and given them the strength to remain true to their purpose in spite of countless failures. It is cosmic religious feeling that gives a man such strength. A contemporary has said, not unjustly, that in this materialistic age of ours the serious workers are the only profoundly religious people.

12

Science and Religion

I

DURING THE LAST CENTURY, and part of the one before, it was widely held that there was an unreconcilable conflict between knowledge and belief. The opinion prevailed among advanced minds that it was time that belief should be replaced increasingly by knowledge; belief that did not itself rest on knowledge was superstition and, as such, had to be opposed. According to this conception, the sole function of education was to open the way to thinking and knowing, and the school, as the outstanding organ for the people's education, must serve that end exclusively.

One will probably find but rarely, if at all, the rationalistic standpoint expressed in such crass form; for any sensible man would see at once how one-sided is such a statement of the position. But it is just as well to state a thesis starkly and nakedly, if one wants to clear up one's mind as to its nature.

It is true that convictions can best be supported with experience and clear thinking. On this point, one must agree unreservedly with the extreme rationalist. The weak point of his conception is, however, this, that those convictions which are necessary and determinant for our conduct and judgments cannot be found solely along this solid scientific way.

For the scientific method can teach us nothing else beyond how facts are related to, and conditioned by, each other. The aspiration toward such objective knowledge belongs to the highest of which man is capable, and you will certainly not suspect me of wishing to belittle the achievements and the heroic efforts of man in this sphere. Yet it is equally clear that knowledge of what *is* does not open the door directly to what *should be.* One can have the clearest and most complete knowledge of what *is,* and yet not be able to deduct from that what should be the *goal* of our human aspirations. Objective knowledge provides us with powerful instruments for the achievements of certain ends, but the ultimate goal itself and the longing to reach it must come from another source. And it is hardly necessary to argue for the view that our existence and our activity acquire meaning only by the setting up of such a goal and of corresponding values. The knowledge of truth as such is wonderful, but it is so little capable of acting as a guide that it cannot prove even the justification and the value of the aspiration toward that very knowledge of truth. Here we face, therefore, the limits of the purely rational conception of our existence.

But it must not be assumed that intelligent thinking can play no part in the formation of the goal and of ethical judgments. When someone realizes that for the achievement of an end certain means would be useful, the means itself becomes thereby an end. Intelligence makes clear to us the interrelation of means and ends. But mere thinking cannot give us a sense of the ultimate and fundamental ends. To make clear these fundamental ends and valuations, and to set them fast in the emotional life of the individual, seems to me precisely the most important function which religion has to perform in the social life of man. And if one asks whence derives the authority of such fundamental ends, since they cannot be stated and justified merely by reason, one can only answer: they exist in a healthy society as powerful traditions, which act upon the conduct and aspirations and judgments of the individuals; they are there, that is, as something living, without its being necessary to find justification for their existence. They come into being not through demonstration but through revelation, through the medium of powerful personalities. One must not attempt to justify them, but, rather, to sense their nature simply and clearly.

The highest principles for our aspirations and judgments are given to us in the Jewish-Christian religious tradition. It is a very high goal which, with our weak powers, we can reach only very inadequately, but which gives a sure foundation to our aspirations and valuations. If one were to take that goal out of its religious form and look merely at its purely human side, one might state it perhaps thus: free and responsible development of the individual, so that he may place his powers freely and gladly in the service of all mankind.

There is no room in this for the divinization of a nation, of a class, let alone of an individual. Are we not all children of one father, as it is said in religious language? Indeed, even the divinization of humanity, as an abstract totality, would not be in the spirit of that ideal. It is only to the individual that a soul is given. And the high destiny of the individual is to serve rather than to rule, or to impose himself in any other way.

If one looks at the substance rather than at the form, then one can take these words as expressing also the fundamental democratic position. The true democrat can worship his nation as little as can the man who is religious, in our sense of the term.

If one holds these high principles clearly before one's eyes, and compares them with the life and spirit of our times, then it appears glaringly that civilized mankind finds itself at present in grave danger. In the totalitarian states, it is the rulers themselves who strive actually to destroy that spirit of humanity. In less threatened parts, it is nationalism and intolerance, as well as the oppression of the individuals by economic means, which threaten to choke these most precious traditions.

A realization of how great is the danger is spreading, however, among thinking people, and there is much search for means with which to meet the danger—means in the field of national and international politics, of legislation, or organization in general. Such efforts are, no doubt, greatly needed. Yet the ancients knew something which we seem to have forgotten. All means prove but a blunt instrument if they have not behind them a living spirit. But if the longing for the achievement of the goal is powerfully alive within us, then shall we not lack the strength to find the means for reaching the goal and for translating it into deeds.

II

It would not be difficult to come to an agreement as to what we understand by science. Science is the century-old endeavor to bring together by means of systematic thought the perceptible phenomena of this world into as thorough-going an association as possible. To put it boldly, it is the attempt at the posterior reconstruction of existence by the process of conceptualization. But when asking myself what religion is, I cannot think of the answer so easily. And even after finding an answer which may satisfy me at this particular moment, I still remain convinced that I can never, under any circumstances, bring together, event to a slight extent, the thoughts of all those who have given this question serious consideration.

At first, then, instead of asking what religion is, I should prefer to ask what characterizes the aspirations of a person who gives me the impression of being religious: a person who is religiously enlightened appears to me to be one who has, to the best of his ability, liberated himself from the fetters of his selfish desires and is preoccupied with thoughts, feelings, and aspirations to which he clings because of their superpersonal value. It seems to me that what is important is the force of this superpersonal content and the depth of the conviction concerning its overpowering meaningfulness, regardless of whether any attempt is made to unite this content with a divine Being, for, otherwise, it would not be possible to count Buddha and Spinoza as religious personalities. Accordingly, *a religious person is devout in the sense that he has no doubt of the significance and loftiness of those superpersonal objects and goals which neither require nor are capable of rational foundation.* They exist with the same necessity and matter-of-factness as he himself. In this sense, religion is the age-old endeavor of mankind to become clearly and completely conscious of these values and goals and constantly to strengthen and extend their effect. If one conceives of religion and science according to these definitions then a conflict between them appears impossible. For science can only ascertain what *is,* but not what *should be,* and outside of its domain value judgments of all kinds remain necessary. Religion, on the other hand, deals only with evaluations of human thought and action: it cannot justifiably speak of facts and relationships between facts. According to this interpretation, the well-known conflicts between religion and science in the past must all be ascribed to a misapprehension of the situation which has been described.

For example, a conflict arises when a religious community insists on the absolute truthfulness of all statements recorded in the Bible. This means an intervention on the part of religion into the sphere of science; this is where the struggle of the Church against the doctrines of Galileo and Darwin belongs. On the other hand, representatives of science have often made an attempt to arrive at fundamental judgments with respect to values and ends on the basis of scientific method, and in this way have set themselves in opposition to religion. These conflicts have all sprung from fatal errors.

Now, even though the realms of religion and science in themselves are clearly marked off from each other, nevertheless there exist between the two strong reciprocal relationships and dependencies. Though religion may be that which determines the goal, it has, nevertheless, learned from science, in the broadest sense, what means will contribute to the attainment of the goals it has set up. But science can only be created by those who are thoroughly imbued with the aspiration toward truth and understanding. This source of feeling, however, springs from the sphere of religion. To this there also belongs the faith in the possibility that the regulations valid for the world of existence are rational, that is, comprehensible to reason. I cannot conceive of a genuine scientist without that profound faith. The situation may be expressed by an image: science without religion is lame, religion without science is blind.

Though I have asserted above that, in truth, a legitimate conflict between religion and science cannot exist, I must nevertheless qualify this assertion once again on an essential point, with reference to the actual content of historical religions. This qualification has to do with the concept of God. During the youthful period of mankind's spiritual evolution human fantasy created gods in man's own image, who, by the operations of their will were supposed to determine or, at any rate, to influence the phenomenal world. Man sought to alter the disposition of these gods in his own favor by means of magic and prayer. The idea of God in the religions taught at present is a sublimation of that old concept of the gods. Its anthropomorphic character is shown, for instance, by the fact that men appeal to the Divine Being in prayers and plead for the fulfillment of their wishes.

Nobody, certainly, will deny that the idea of the existence of an omnipotent, just, and omnibeneficent personal God is able to accord man solace, help, and guidance; also, by virtue of its simplicity, it is accessible to the most undeveloped mind. But, on the other hand, there are decisive weaknesses attached to this idea in itself, which have been painfully felt since the beginning of history. That is, if this being is omnipotent, then every occurrence—including every human action, every human thought, and every human feeling and aspiration—is also His work; how is it possible to think of holding men responsible for their deeds and thoughts before such an almighty Being? In giving out punishment and rewards, He would, to a certain extent, be passing judgment on Himself. How can this be combined with the goodness and righteousness ascribed to Him?

The main source of the present day conflicts between the spheres of religion and of science lies in this concept of a personal God. It is the aim of science to establish general rules which determine the reciprocal connection of objects and events in time and space. For these rules, or laws of nature, absolutely general validity is required—not proven. It is mainly a program, and faith in the possibility of its accomplishment in principle is only founded on partial successes. But hardly anyone could be found who would deny these partial successes and ascribe them to human self-deception. The fact that, on the basis of such laws, we are able to predict the temporal behavior of phenomena in certain domains with great precision and certainty is deeply embedded in the consciousness of the modern man, even though he may have grasped very little of the contents of those laws. He need only consider that planetary courses within the solar system may be calculated in advance with great exactitude on the basis of a limited number of simple laws. In a similar way, though not with the same precision, it is possible to calculate in advance the mode of operation of an electric motor, a transmission system, or of a wireless apparatus, even when dealing with a novel development.

To be sure, when the number of factors coming into play in a phenomenological complex is too large, scientific method in most cases fails us. One need only think of the weather, in which case prediction even for a few days ahead is impossible. Nevertheless, no one doubts that we are confronted with a causal connection whose causal components are in the main known to us. Occurrences in this domain are beyond the reach of exact prediction because of the variety of factors in operation, not because of any lack of order in nature.

We have penetrated far less deeply into the regularities obtaining within the realm of living things, but deeply enough, nevertheless, to sense at least the rule of fixed necessity. One need only think of the systematic order in heredity, and in the effect of poisons, as, for instance, alcohol, on the behavior of organic beings. What is still lacking here is a grasp of connections of profound generality, but not a knowledge of order in itself.

The more a man is imbued with the ordered regularity of all events, the firmer becomes his conviction that there is no room left by the side of this ordered regularity for causes of a different nature. For him, neither the rule of human nor the rule of divine will exists as an independent cause of natural events. To be sure, the doctrine of a personal God interfering with natural events could never be *refuted*, in the real sense, by science, for this doctrine can always take refuge in those domains in which scientific knowledge has not yet been able to set foot.

But I am persuaded that such behavior on the part of the representatives of religion would not only be unworthy but also fatal. For a doctrine which is able to maintain itself not in clear light but only in the dark, will, of necessity, lose its effect on mankind, with incalculable harm to human progress. In their struggle for the ethical good, teachers of religion must have the stature to give up the doctrine of a personal God, that is, give up that source of fear and hope which in the past placed such vast power in the hands of priests. In their labors, they will have to avail themselves of those forces which are capable of cultivating the Good, the True, and the Beautiful in humanity itself. This is, to be sure, a more difficult but an incomparably more worthy task. After religious teachers accomplish the refining process indicated, they will surely recognize with joy that true religion has been ennobled and made more profound by scientific knowledge.

If it is one of the goals of religion to liberate mankind as far as possible from the bondage of egocentric cravings, desires, and fears, scientific reasoning can aid religion in yet another sense. Although it is true that it is the goal of science to discover rules which permit the association and foretelling of facts, this is not its only aim. It also seeks to reduce the connections discovered to the smallest possible number of mutually independent conceptual elements. It is in this striving after the rational unification of the manifold that it encounters its greatest successes, even though it is precisely this attempt which causes it to run the greatest risk of falling a prey to illusions. But whoever has undergone the intense experience of successful advances made in this domain is moved by profound reverence for the rationality made manifest in existence. By way of the understanding he achieves a far-reaching emancipation from the shackles of personal hopes and desires, and thereby attains that humble attitude of mind toward the grandeur of reason incarnate in existence, and which, in its profoundest depths, is inaccessible to man. This attitude, however, appears to me to be religious in the highest sense of the word. And so it seems to me that science not only purifies the religious impulse of the dross of its anthropomorphism, but also contributes to a religious spiritualization of our understanding of life.

The interpretation of religion, as here advanced, implies a dependence of science on the religious attitude, a relation which, in our predominantly materialistic age, is only too easily overlooked. While it is true that scientific results are entirely independent from religious or moral considerations, those individuals to whom we owe the great creative achievements of science were all of them imbued with the truly religious conviction that this universe of ours is something perfect and susceptible to the rational striving for knowledge. If this conviction had not been a strongly emotional one and if those searching for knowledge had not been inspired by Spinoza's *Amor Dei Intellectualis,* they would hardly have been capable of that untiring devotion which alone enables man to attain his greatest achievements.

DE BROGLIE

PRINCE LOUIS DE BROGLIE
(1892–1987)

L OUIS DE BROGLIE is best known for his theory of "matter waves," the crucial formulations of which he presented in two papers of September 1923, while he was still a student. These papers became part of his doctoral thesis, a copy of which was sent to Einstein, who, much impressed, widely circulated the ideas. Erwin Schroedinger heard of de Broglie's thesis—that moving electrons produce waves—and that directly led him to develop the Schroedinger wave equations so central to quantum mechanics. The actual existence of matter waves was experimentally verified in 1927, and two years later de Broglie received the Nobel Prize in Physics.

The following sections are taken from *Physics and Microphysics* (New York: Pantheon, 1955). In the first section, de Broglie argues (as did Einstein) that all genuine science is motivated by what, in fact, are spiritual ideals. But science itself cannot pronounce on these ideals, and thus, in the second section, he argues that, in addition to science, we need "a supplement of the soul."

13

The Aspiration Towards Spirit

MAN EXPERIENCES the desire and almost the need for knowledge. From the most distant origins of history we see him preoccupied with the study of the phenomena of the world which surrounds him and with the endeavour to explain them. Assuredly, these primitive explanations seem very naïve to us, all impregnated as they are with mythology and anthropocentrism. They are, nonetheless, the first signs of the curiosity and anxiety which lead the human mind to attempt to understand and coordinate the facts that he observes in nature; they are the first affirmations of that bold act of faith which leads us to bear witness to the existence of a certain correlation between the succession of natural phenomena on the one hand, and the pictures or reasonings which our mind is able to conceive on the other hand.

As our attainments were freed from the mists in which they were at first immersed, science took its modern form. Thus scientists have come to feel more and more keenly that there exists in nature an order, a harmony, which is at least partially accessible to our intelligence, and they have devoted all their efforts to discover each day more of the nature and the extent of this harmony. Thus was born what we often call "pure science," that is, that activity of our mind which has as its goal the knowledge of natural phenomena and of establishing amongst them rational relations, independently of all utilitarian preoccupation. At the same time, and as an addition, by teaching us more about the laws which govern phenomena, the development of science has progressively allowed for a great number of inventions and practical applications which have completely transformed, often for good and sometimes for evil, the living conditions of humanity.

In the presence of the immense effort which humanity has exerted for generations past, and which it incessantly develops in our day, to succeed in extending and promoting disinterested knowledge of natural phenomena and of their coordination, one question is forced on our attention. What, in fact, is the *raison d'être* of this effort, what mysterious attraction acting on certain men urges them to dedicate their time and labours to works from which they themselves often hardly profit? How, for the unique pleasure of obtaining a momentary glimpse of some new aspect of truth, in the midst of the besetting preoccupations of daily life, in the midst of the conflict of interests of which it all consists, has pure science, single-mindedly turned toward the ideal, been able to find its way? Evidently, this is one of the aspects of the dual nature of man, so often put into relief by thinkers and philosophers; restrained by our organic constitution and by our different emotions in the lower sphere of our daily occupations, we also feel ourselves urged on by the appeal of the ideal, by more or less precise aspiration towards spiritual values, and from those sentiments even the worst amongst us do not entirely escape. But this general explanation of high aspirations and disinterested efforts through man's moral nature, which are applied to so many different realms of human activity, is not in itself completely sufficient to account for the attraction which pure and disinterested science exerts on our mind. We must try to define still further the origin and nature of this attraction.

What, then, is the goal pursued, sometimes without being clearly aware of it, by the experimenter who works in his laboratory to determine the nature of the known phenomena or to observe new ones, and the theorist who, in his study, seeks to combine symbols and numbers to draw from them abstract constructions, establishing amongst the observable facts correlations or unsuspected resemblances? This goal, as we have seen, is, without doubt, to succeed in penetrating further into the knowledge of natural harmonies, to come to have a glimpse of a reflection of the order which rules in the universe, some portions of the deep and hidden realities which constitute it. Even the scientists or philosophers who, pragmatic in tendency, have reduced to a utilitarian role the value of scientific theories, as, for example, the eminent physicist Pierre Duhem, have had to recognize that these theories establish between the phenomena a "natural classification," allowing us to sense the existence of an "ontological order" which is beyond us. All those who dedicate their efforts to pure science admit, whether they agree with it or not, the existence of such an order and it is to enable them to lift up for a moment, from one distant point to the next, a corner of the veil which conceals it from us that they expend their strength and their vigils. The great epoch-making discoveries in the history of science (think, for example, of that of universal gravitation) have been like sudden lightning flashes, making us perceive in one single glance a harmony up till then unsuspected, and it is to have, from time to time, the divine joy of discovering such harmonies that pure science works without sparing its toil or seeking for profit.

Assuredly, the great discoveries are not accomplished in a day; it is necessary that they should be prepared for a long time by meticulous and austere labours. At times, immersed in the details of his absorbing work, the specialist, preparing his apparatus or developing his calculations, can very well lose sight of the far-off goal of his researches and be no longer greatly concerned with the harmony of the universe; yet, what gives value to his efforts, what justifies its apparent uselessness, is that he thus supplies to the common task small contributions capable one day of facilitating the erection of some of those great syntheses which do honour to the human mind. Pure science untiringly pursues the search for this hidden order, these ultimate realities; each scientist conceives their existence and significance in his own way, according to the inclinations or philosophical convictions which influence him, but all scientists, when they are sincere, recognize that the search for truth is the real reason that justifies the efforts of pure science and constitutes its nobility. Moreover, on this important question of the goal of disinterested science, all true scientists, in spite of the differences of opinion which can separate them, are, without doubt, nearer to being in agreement than they themselves often imagine.

The great wonder in the progress of science is that it has revealed to us a certain agreement between our thought and things, a certain possibility of grasping, with the assistance of the resources of our intelligence and the rules of our reason, the profound relations existing between phenomena. We are not sufficiently astonished by the fact that any science may be possible, that is, that our reason should provide us with the means of understanding at least certain aspects of what happens around us in nature. Some thinkers, nevertheless, find this fact natural because, they say, humanity having had to endure during thousands of years the consequences of natural phenomena and to learn, in order to survive, to adapt itself to it, our mind has thus learned little by little to form its logic and its rules of reasoning under the pressure of the material world, and it must not, in consequence, be astonished to recover in the material world the logic and the rules of reasoning that it has extracted from it.

Personally, we do not find this argument very conclusive; in reality, in order that humanity should have been able to adapt itself to live in the world which surrounds us, it would undoubtedly be necessary that there should be already between this world and our mind some analogy in structure; if that had not been so, perhaps humanity would not have been able to survive. Well, it would have disappeared, that is all! Since it has survived, it is then, because it was capable of understanding certain of the rules which govern the succession of natural phenomena, in a way to adapt itself to these phenomena or even to utilize them to its advantage. This is why the pre-adaptation of our mind to the discovery of relations between phenomena and the order that is manifested in nature appears to us much more surprising than it is sometimes said to be.

What appears to us to show well that we can hardly explain this pre-adaptation by a secular experience dating from the origins of humanity, is as follows. In several cases, especially in the most recent science, the minute study of phenomena very delicate to observe, a study very different from the rough experiments that the caveman was able to make, has led us to discover in the depths of our own mind hitherto unsuspected resources, allowing us to interpret our new discoveries and to give to them an intelligible meaning. In saying this, we are thinking especially of the remarkable new theories of contemporary physics. Take, for instance, the theory of relativity; starting from extremely delicate and precise experiments, the results of which could not be foreseen by the older theories, it built up a new conception of space and time and of their reciprocal relations, a conception absolutely contrary to all the data of our usual intuition; it thus shows us that our mind can find in itself the necessary elements logically to constitute an interpretation of the ideas of space and time quite different from that which the experience of daily life suggests. By its successes, the theory of relativity therefore shows us how extensive is the parallelism which exists between the rules of our reasoning and the order which conceals itself behind the subtle phenomena which physics of today studies; it shows us that this parallelism infinitely surpasses all that the daily experience of the older generations was able to suggest to us.

More remarkable still is the example which can be drawn from physics of the atomic or microscopic scale, where the theory of quanta and of its extensions rules today. Here, still more so than in the case of the theory of relativity, we have had to appeal to conceptions very far removed from those which we have been accustomed to handle. To account for the phenomena of the atomic scale, we have been obliged, little by little, to abandon the idea that the movement of a corpuscle can be represented by a continuous succession of positions in space, by a trajectory progressively described with a certain specified speed. We have also had to abandon the traditional idea that phenomena, even elementary ones, are rigorously determined and exactly predictable, and to substitute for the rigid determinism of classical physics a more flexible conception, admitting that there exist at each instant in the evolution of elementary phenomena verifiable by us different eventualities concerning which it is only possible to estimate the relative probabilities. We had, in addition, to abandon also all our intuitive and customary ideas on the individuality of corpuscles, on the role of the constituents in a complex system, etc. . . . . In an account like this, it is not possible for us to dwell upon the detail of these difficult questions, but it seems to us essential to make the following remark.

In the development of these theories so daringly novel, which have been, let it be emphasized, imposed on us by the discovery of certain experimental facts, it has been possible to construct on the basis of these new conceptions a perfectly logical formalism, perfectly consistent with the rules of our reason, which allow of the assembling and connecting amongst themselves of all the ascertainable facts in the atomic scale. Here again, we have found in our mind all the resources necessary to represent the order which rules in the atomic scale, although this order is stupendously different from what our imagination could conceive by starting from the usual perceptivity. And this fact seems to us sufficiently independent of the distant past of humanity.

In short, all these examples show us how remarkable is the harmony between the resources of which our mind disposes and the profound realities which conceal themselves behind natural appearances. To bring this harmony more completely into the light, to glimpse yet more the ontological order of which Duhem spoke—such appears to be the true mission of pure science. Removed from all utilitarian preoccupation, solely devoted to the search for truth, it appears to us as one of the noblest activities of which we are capable. By the wholly ideal nature of the goal it pursues, by the intensity and the disinterested character of the efforts that it demands, it possesses a moral value which cannot be denied.

Perhaps we may ask ourselves where this passionate quest for truth can lead us. Science advances with great strides; astonishing discoveries issue each day from its laboratories; by its bold theories it opens out for us wonderful new vistas on the mystery of things. Will it then soon lift the veil of Isis, make us definitely penetrate into the secrets of nature, give an assured answer to the great metaphysical problems which for so many centuries vex the soul of man? It does not seem that we are yet near the attainment of such a triumph of pure science. Mystery surrounds us; as Puvis of Chavannes has symbolically represented it in the vast fresco which adorns the great amphitheatre of the Sorbonne, we are placed as at the centre of a small clearing surrounded on all sides by an immense and gloomy, unexplored forest. No, it is not yet tomorrow that science will be able to give us the key to the enigmas of the universe; we are not yet near to the attainment of the end of an effort to which nothing permits us to fix the duration.

Nevertheless, it is not impossible that the advances of science will bring new data capable, if not of solving, at least of clarifying certain great problems of philosophy. Already contemporary physics, by introducing its new ideas on space and time, on the impossibility of following the determinism of elementary phenomena, on the "complementary" character of certain pictures apparently contradictory (such as that of wave and of corpuscle), on the inability to discern elementary particles, already contemporary physics, I say, offers to the meditations of philosophical minds entirely new themes of which, at the present hour, we are far from having perceived all the consequences. The study of the nucleus of atoms, by making us penetrate to the extreme depths of matter, reserves many surprises for us and may bring us important revelations. Astronomy, by extending in an unheard of way the limits of the observable region of the stellar world, already brings to us data about the extent of the universe, its age, its evolution, calculated to orientate our cosmological conceptions. Still many other sciences furnish us each day with similar information concerning which the thoughts of the philosophers of the future will have to take account. But it is necessary to reserve a special place for biology; this is a science of capital importance because it is the science of life. Its advances are rapid; its discoveries, especially in genetics, are of captivating interest. Perhaps it will bring us, sooner or later, very important indications as to the role of the phenomena of life and the real place which it is suitable to attribute to them in the whole of nature.

Thus, therefore, science progresses; it marches on and will no doubt each day march further forward on the road to a better comprehension of natural phenomena. From this point of view, all hopes are permissible and the human mind will undoubtedly gather the fruits of its ceaseless secular efforts to discover new facets of the truth. The beauty and the moral grandeur of pure science, the progress that it achieves, the joy of knowing that it is deserving of the enthusiasm that it inspires in its adepts!

Alas! Why is it necessary that there should be shadows cast over this bright picture? Why must it unfortunately be that the applications of science should not necessarily be beneficial? How could we forget it in an epoch when on all sides there sounds the tumult of arms and where there accumulate the ruins caused by the terrifying new means of destruction? If, through the progress of science, new possibilities for the amelioration of the lot of mankind are offered us, it is only too certain that at the very same time powerful new means for causing suffering, for killing and destroying, will be placed at our disposal. Shall we be wise enough not to make use of them, or, at least, not to abuse them? And, since we are giving free scope to our imagination, we can also suppose that, in a future more or less distant, the progress of genetics will permit us to call forth the appearance of new types of living beings who might be supermen, but also monsters; how, endowed with such power, would men make use of it? Would humanity be wise enough not to employ the new arms, which science would have provided, to perpetrate its own destruction?

At bottom, these distressing questions raise, above all, a moral problem. Scientific discoveries and the applications of which they are capable are, in themselves, neither good nor bad; all depends on the use which we make of them. Tomorrow, as today, it will be, therefore, the will of mankind that is called upon to decide on the beneficial or evil character of these applications. To be able to survive the appropriate progress of his attainments, mankind of tomorrow will have to find in the development of his spiritual life and in the uplifting of his moral ideal, the wisdom not to abuse his increased forces. This is what Henri Bergson has splendidly expressed in one of his last works when saying: "Our enlarged body clamours for an addition to the spirit." Shall we be able to acquire this addition to the spirit as rapidly as the advances of science will develop?

14

The Mechanism Demands a Mysticism

I N THE LAST CHAPTER OF HIS GREAT WORK, *The Two Sources of Morality and Religion,* Henri Bergson, having reached almost the end of his book, showed to us a humanity in the formidable grip of mechanism, and as if succumbing under the weight of the discoveries and inventions which the creative activity of its mind had been able to realize. Doubtless it was as though inscribed in the destinies of man, of *homo sapiens,* of *homo faber,* some day to use the forces of nature for his advantage, but Bergson rightly said:

. . . machines which move on petrol, on coal, or hydro-electric power and which convert into motion the potential energies accumulated during millions of years, have given to our organism so vast an extension and so formidable a power, so disproportionate to its dimensions and strength, that surely it had never been foreseen in the plan of the structure of our species.

And wishing to make us appreciate the essential point and the disquieting side of the problem, he added: "Now, in this excessively enlarged body, the spirit remains what it was, too small now to fill it, too feeble to direct it," and further on, "Let us add that this increased body awaits a supplement of the soul and that the mechanism demands a mysticism." Finally, the work finishes on these words, pregnant with meaning: "Humanity groans half-crushed under the weight of the advances that it has made. It does not know sufficiently that its future depends on itself. It is for it, above all, to make up its mind if it wishes to continue to live. . . ."

Today, on the morrow of the discovery of the atom bomb which has shown that, henceforth, man can at will utilize those formidable reserves of energy which are concealed in the very heart of matter—the nuclei of atoms—we understand much better the terrifying extent of the anxieties which the distinguished author of *Creative Evolution* expressed some twenty years ago. We shall realize to what extent this new source of energy is superior to all those which we previously were able to utilize, if I recall that, in the phenomena of the fission of uranium, a single gramme of uranium can furnish us with as much energy as more than ten tons of coal, and it is not out of order to think that in the utilization of atomic energy we shall attain still greater returns. Thereby we shall acquire a new power of which it is almost impossible to fix the limits. But it must be stated with some sorrow that it was in order to produce an engine of formidable power of destruction that men, for the first time, made use of this fundamentally admirable discovery of modern physics; this fact, in itself is, somewhat disquieting.

Without doubt, atomic energy will very probably be capable of beneficent applications. It will allow us to economize our reserves of coal and petrol, make our factories function, turn our motors; transformed, for example, into electric energy, it will supply us in almost unlimited quantity with motive power, heat, and light. The era of atomic energy can be an era of admirable progress, an era of a better and easier life. But it can also be an era of inexpiable strife, surpassing in extent and horror all the wars of the past where, with the aid of terrifying means of destruction, humanity runs the risk of completely destroying itself. It would serve no purpose if we deceived ourselves by misleading illusions on the possibility of such catastrophes for, alas! human passions have remained the same, and the terrible events which, in these last years, have stained all parts of the world with blood scarcely allow us to hope that, in the future, the wisdom and love of his fellow man will necessarily always prevail in the heart of men.

And so the drama presents itself in the sense that Bergson had foreseen. With our power of action suddenly and enormously increased, will our enfeebled spirit, nevertheless, be sufficiently strong to put it to good use? It is on our will and our will alone that there is going to depend the good or the evil use of the unheard of forces which science has handed over to our control. We perceive the almost tragic magnitude of the moral problem which is here raised. "Humanity does not know sufficiently that its future depends on itself. It is for it to see first if it wishes to continue to live," said Bergson. How precise and profound a meaning these words hold today on the threshold of the unknown, and perhaps formidable, future which opens before us!

So long as the means available to man, which enabled him to act on his surroundings, were limited, the consequences of his evil actions were themselves also limited, and the damage which resulted from them for human societies was able to be expunged more or less rapidly. In proportion, as the means of action and, consequently, of destruction, placed at our disposal by the progress of the sciences and of technics, were developed, the ravage that they were capable of producing has become more and more extended and the wounds thus produced, because they were deeper, have taken longer to heal. The evolution of wars in modern times offers a tragic example of this. They have involved more and more extensive regions of the earth, casting into the furnace of battle a larger and larger number of combatants, each day exposing more of the civilian populations to the same dangers as the soldiers. With the use of atomic energy, the wars of tomorrow can assume an infernal character, the whole horror of which it is difficult to imagine.

But it is not only wars that could have terrible consequences in the future. We can easily imagine other catastrophes which show still better the nature of the danger. Fifty years ago some heated people, whom we then called anarchists, threw bombs in different public places, into cafés, churches, and even into the Chamber of Deputies. These attempts caused the death of a certain number of people and justly roused public indignation. The number of victims and of material destruction was, however, limited. How much greater would it be if tomorrow new anarchists succeeded in utilizing, in their criminal attempts, atomic bombs capable of destroying entire cities! Here we really grasp the true nature of the drama; the will of one man only can become sufficient to unleash a phenomenon of a formidable power. By this enormous increase in power the responsibility of man is augmented in a like proportion, and the consequences of a moral weakening can become incalculable. Hence the moral problem acquires a significance much greater than in the past. It seems, to use Bergson's language, that our souls have not grown in proportion to our bodies, and, therefore, humanity will not have any excess of the spiritual forces on which morale can lean without too much peril in following the dangerous roads ahead.

I have just set forth several somewhat engrossing aspects of the present evolution of scientific progress. Nevertheless, it is not necessary to lose all faith in the future. Humanity has already passed through many difficult situations and it has always succeeded in emerging to its advantage. Often the feeling of the imminence of a danger gives birth in the heart of men to sentiments or mysticisms which can serve to avoid it. It is necessary that we should thoroughly realize the peril that the bad use of atomic energy would constitute for our species in order to promote in us the reactions which can preserve us from danger. It is somewhat comforting from this point of view to aver that in the war which has just ended, though so terrible in so many respects, none of the adversaries dared to make use of toxic gases. It may be due to a feeling of moral responsibility or fear of reprisals, perhaps both, but the fact is there; toxic gases were not used.

Confronted by the dangers with which the advances of science can, if employed for evil, face him, man has need of a "supplement of soul" and he must force himself to acquire it promptly before it is too late. It is the duty of those who have the mission of being the spiritual or intellectual guides of humanity to labour to awaken in it this supplement of the soul.

JEANS

SIR JAMES JEANS
(1877–1946)

S IR JAMES JEANS was a mathematician, physicist, and astronomer. He made fundamental contributions to the dynamical theory of gases, the mathematical theory of electromagnetism, the evolution of gaseous stars, the nature of nebulae—to name a few. He was knighted in 1924 and went on to become one of the most popular and prominent philosophers of science.

The following is taken from *The Mysterious Universe* (Cambridge University Press, 1931). Sir Jeans concludes that, since we can only understand the physical world through mathematics, then we might rightly conclude that, to use his favorite phrase, "God is a mathematician, and the universe begins to look more like a great thought than a great machine." He makes it very clear he is talking now as a philosopher, not a scientist, but his Pythagorean mysticism inspires a style that manages to embrace both with delight, rigor, and wit.

15

In the Mind of Some Eternal Spirit

THE ESSENTIAL FACT is simply that *all* the pictures which science now draws of nature, and which alone seem capable of according with observational fact, are *mathematical* pictures.

Most scientists would agree that they are nothing more than pictures—fictions, if you like, if by fiction you mean that science is not yet in contact with ultimate reality. Many would hold that, from the broad philosophical standpoint, the outstanding achievement of twentieth-century physics is not the theory of relativity with its welding together of space and time, or the theory of quanta with its present apparent negation of the laws of causation, or the dissection of the atom with the resultant discovery that things are not what they seem; it is the general recognition that we are not yet in contact with ultimate reality. To speak in terms of Plato's well-known simile, we are still imprisoned in our cave, with our backs to the light, and can only watch the shadows on the wall. At present, the only task immediately before science is to study these shadows, to classify them and explain them in the simplest possible way. And what we are finding, in a whole torrent of surprising new knowledge, is that the way which explains them more clearly, more fully, and more naturally than any other is the mathematical way, the explanation in terms of mathematical concepts. It is true, in a sense somewhat different from that intended by Galileo, that "Nature's great book is written in mathematical language." So true is it that no one, except a mathematician, need ever hope fully to understand those branches of science which try to unravel the fundamental nature of the universe—the theory of relativity, the theory of quanta, and the wave-mechanics.

The shadows which reality throws onto the wall of our cave might *a*

135

priori have been of many kinds. They might conceivably have been perfectly meaningless to us, as meaningless as a cinematograph film showing the growth of microscopic tissues would be to a dog who had strayed into a lecture room by mistake. Indeed, our earth is so infinitesimal in comparison with the whole universe, we, the only thinking beings, so far as we know, in the whole of space, are, to all appearances, so accidental, so far removed from the main scheme of the universe, that it is *a priori* all too probable that any meaning that the universe as a whole may have, would entirely transcend our terrestrial experience and so be totally unintelligible to us. In this event, we should have had no foothold from which to start our exploration of the true meaning of the universe.

Although this is the most likely event, it is not impossible that some of the shadows thrown onto the walls of our cave might suggest objects and operations with which we cave-dwellers were already familiar in our caves. The shadow of a falling body behaves like a falling body, and so would remind us of bodies we had ourselves let fall; we should be tempted to interpret such shadows in mechanical terms. This explains the mechanical physics of the last century; the shadows reminded our scientific predecessors of the behaviour of jellies, spinning-tops, thrust-bars, and cog-wheels, so that they, mistaking the shadow for the substance, believed they saw before them a universe of jellies and mechanical devices. We know now that the interpretation is conspicuously inadequate: it fails to explain the simplest phenomena, the propagation of a sunbeam, the composition of radiation, the fall of an apple, or the whirl of electrons in the atom.

Again, the shadow of a game of chess, played by the actors out in the sunlight, would remind us of the games of chess we had played in our cave. Now and then we might recognize knights' moves, or observe castles moving simultaneously with kings and queens, or discern other characteristic moves so similar to those we were accustomed to play that they could not be attributed to chance. We would no longer think of the external reality as a machine; the details of its operation might be mechanical, but, in essence, it would be a reality of thought: we should recognize the chess players out in the sunlight as beings governed by minds like our own; we should find the counterpart of our own thoughts in the reality which was forever inaccessible to our direct observation.

And when scientists study the world of phenomena, the shadows which nature throws onto the wall of our cave, they do not find these shadows totally unintelligible, and neither do they seem to represent unknown or unfamiliar objects. Rather, it seems to me, we can recognize chess players outside in the sunshine who appear to be very well acquainted with the rules of the game *as we have formulated them in our cave.* To drop our metaphor, nature seems very conversant with the rules of pure mathematics as our mathematicians have formulated them in their studies, out of their own inner consciousness and without drawing to any appreciable extent on their experience of the outer world. By "pure mathematics" is meant those departments of mathematics which are creations of pure thought, of reason operating solely within her own sphere, as contrasted with "applied mathematics" which reasons about the external world, after first taking some supposed property of the external world as its raw material. Descartes, looking round for an example of the produce of pure thought uncontaminated by observation (rationalism), chose the fact that the sum of the three angles of a triangle was necessarily equal to two right angles. It was, as we now know, a singularly unfortunate choice. Other choices, far less open to objection, might easily have been made, as, for instance, the laws of probability, the rules of manipulation of "imaginary" numbers—i.e., numbers containing the square roots of negative quantities—or multi-dimensional geometry. All these branches of mathematics were originally worked out by the mathematician in terms of abstract thought, practically uninfluenced by contact with the outer world, and drawing nothing from experience: they formed

an independent world
created out of pure intelligence.

And now it emerges that the shadow-play which we describe as the fall of an apple to the ground, the ebb and flow of the tides, the motion of electrons in the atom, are produced by actors who seem very conversant with these purely mathematical concepts—with our rules of our game of chess, which we formulated long before we discovered that the shadows on the wall were also playing chess.

When we try to discover the nature of the reality behind the shadows, we are confronted with the fact that all discussion of the ultimate nature of things must necessarily be barren unless we have some extraneous standards against which to compare them. For this reason, to borrow Locke's phrase, "the real essence of substances" is forever unknowable. We can only progress by discussing the laws which govern the changes of substances, and so produce the phenomena of the external world. These we can compare with the abstract creations of our own minds.

For instance, a deaf engineer studying the action of a pianola might try first to interpret it as a machine, but would be baffled by the continuous reiteration of the intervals 1, 5, 8, 13 in the motions of its trackers. A deaf musician, although he could hear nothing, would immediately recognize this succession of numbers as the intervals of the common chord, while other successions of less frequent occurrence would suggest other musical chords. In this way, he would recognize a kinship between his own thoughts and the thoughts which had resulted in the making of the pianola; he would say that it had come into existence through the thought of a musician. In the same way, a scientific study of the action of the universe has suggested a conclusion which may be summed up, though very crudely and quite inadequately, because we have no language at our command except that derived from our terrestrial concepts and experiences, in the statement that the universe appears to have been designed by a pure mathematician.

This statement can hardly hope to escape challenge on the ground that we are merely moulding nature to our preconceived ideas. The musician, it will be said, may be so engrossed in music that he would contrive to interpret every piece of mechanism as a musical instrument; the habit of thinking all intervals as musical intervals may be so ingrained in him that if he fell downstairs and bumped on stairs numbered 1, 5, 8, and 13 he would see music in his fall. In the same way, a cubist painter can see nothing but cubes in the indescribable richness of nature—and the unreality of his pictures shows how far he is from understanding nature; his cubist spectacles are mere blinkers which prevent his seeing more than a minute fraction of the great world around him. So, it may be suggested, the mathematician only sees nature through the mathematical blinkers he has fashioned for himself. We may be reminded that Kant, discussing the various modes of perception by which the human mind apprehends nature, concluded that it is specially prone to see nature through mathematical spectacles. Just as a man wearing blue spectacles would see only a blue world, so Kant thought that, with our mental bias, we tend to see only a mathematical world. Does our argument merely exemplify this old pitfall, if such it is?

A moment's reflection will show that this can hardly be the whole story. The new mathematical interpretation of nature cannot all be in our spectacles—in our subjective way of regarding the external world—

since if it were we should have seen it long ago. The human mind was the same in quality and mode of action a century ago as now; the recent great change in scientific outlook has resulted from a vast advance in scientific knowledge and not from any change in the human mind; we have found something new and hitherto unknown in the objective universe outside ourselves. Our remote ancestors tried to interpret nature in terms of anthropomorphic concepts of their own creation and failed. The efforts of our nearer ancestors to interpret nature on engineering lines proved equally inadequate. Nature refused to accommodate herself to either of these man-made moulds. On the other hand, our efforts to interpret nature in terms of the concepts of pure mathematics have, so far, proved brilliantly successful. It would now seem to be beyond dispute that, in some way, nature is more closely allied to the concepts of pure mathematics than to those of biology or of engineering, and even if the mathematical interpretation is only a third man-made mould, it at least fits objective nature incomparably better than the two previously tried.

A hundred years ago, when scientists were trying to interpret the world mechanically, no wise man came forward to assure them that the mechanical view was bound to prove a misfit in the end—that the phenomenal universe would never make sense until it was projected on to a screen of pure mathematics: had they brought forward a convincing argument to this effect, science might have been saved much fruitless labour. If the philosopher now says, "What you have found is nothing new: I could have told you that it must be so all the time," the scientist may reasonably inquire, "Why, then, did you not tell us so, when we should have found the information of real value?"

Our contention is that the universe now appears to be mathematical in a sense different from any which Kant contemplated or possibly could have contemplated—in brief, the mathematics enters the universe from above instead of from below.

In one sense, it may be argued that everything is mathematical. The simplest form of mathematics is arithmetic, the science of numbers and quantities—and these permeate the whole of life. For instance, commerce, which consists largely of the arithmetical operations of bookkeeping, stock-taking and so on, is, in a sense, a mathematical occupation—but it is not in this sense that the universe now appears to be mathematical.

Again, every engineer has to be something of a mathematician; if he is to calculate and predict the mechanical behaviour of bodies with accuracy, he must use mathematical knowledge and look at his problems through mathematical spectacles—but, again, it is not in this way that science has begun to see the universe as mathematical. The mathematics of the engineer differs from the mathematics of the shopkeeper only in being far more complex. It is still a mere tool for calculation; instead of evaluating stock-in-trade or profits, it evaluates stresses and strains or electric currents.

On the other hand, Plutarch records that Plato used to say that God forever geometrises—Πλάτων ἔλεγε τὸν θεὸν ἀεὶ γεωμετρεῖν—and he sets an imaginary symposium at work to discuss what Plato meant by this. Clearly, he meant something quite different in kind from what we mean when we say that the banker forever arithmetises. Among the illustrations given by Plutarch are: that Plato had said that geometry sets limits to what would otherwise be unlimited, and that he had stated that God had constructed the universe on the basis of the five regular solids—he believed that the particles of earth, air, fire, and water had the shapes of cubes, octahedra, tetrahedra, and icosahedra, while the universe itself was shaped like a dodecahedron. To these may perhaps be added Plato's belief that the distances of the sun, moon, and planets were "in the proportion of the double intervals," by which he meant the sequence of integers which are powers of either 2 or 3—namely 1, 2, 3, 4, 8, 9, 27.

If any of these considerations retain any shred of validity today, it is the first—the universe of the theory of relativity is finite just because it is geometrical. The idea that the four elements and the universe were in any way related to the five regular solids was, of course, mere fancy, and the true distances of the sun, moon, and planets bear absolutely no relation to Plato's numbers.

Two thousand years after Plato, Kepler spent much time and energy in trying to relate the sizes of the planetary orbits to musical intervals and geometrical constructions; perhaps he, too, hoped to discover that the orbits had been arranged by a musician or a geometer. For one brief moment, he believed he had found that the ratios of the orbits were related to the geometry; of the five regular solids. If this supposed fact had been known to Plato, what a proof he might have seen in it of the geometrising propensities of the deity! Kepler himself wrote: "The intense pleasure I have received from this discovery can never be told in words." It need hardly be said that the great discovery was fallacious. Indeed, our modern minds immediately dismiss it as ridiculous; we find it impossible to think of the solar system as a finished product, the same today as when it came from the hand of its maker; we can only think of it as something continually changing and evolving, working out its own future from its past. Yet if we can momentarily give a sufficiently medieval cast to our thoughts and imagine anything so fanciful as the Kepler's conjecture should have been true, it is clear that he would have been entitled to draw some sort of inference from it. The mathematics which he had found in the universe would have been something more than he had himself put in, and he could legitimately have argued that there was inherent in the universe a mathematics additional to that which he had used to unravel its design; he might have argued, in anthropomorphic language, that his discovery suggested that the universe had been designed by a geometer. And he need no more have troubled about the criticism that the mathematics he had discovered resided merely in his own mathematical spectacles, than the angler who catches a big fish by using a little fish as bait need be worried by the comment, "Yes, but I saw you put the fish in yourself."

Let us take a more modern and less fanciful example of the same thing. Fifty years ago, when there was much discussion on the problem of communicating with Mars, it was desired to notify the supposed Martians that thinking beings existed on the planet Earth, but the difficulty was to find a language understood by both parties. The suggestion was made that the most suitable language was that of pure mathematics; it was proposed to light chains of bonfires in the Sahara, to form a diagram illustrating the famous theorem of Pythagoras, that the squares on the two smaller sides of a right-angled triangle are together equal to the square on the greatest side. To most of the inhabitants of Mars such signals would convey no meaning, but it was argued that mathematicians on Mars, if such existed, would surely recognize them as the handiwork of mathematicians on Earth. In so doing, they would not be open to the reproach that they saw mathematics in everything. and it seems to me that the situation is similar, *mutatis mutandis,* with the signals from the outer world of reality which form the shadows on the walls of the cave in which we are imprisoned. We cannot interpret these as shadows cast by living actors nor as shadows cast by a machine, but the pure mathematician recognizes them as representing the kind of ideas with which he is already familiar in his studies.

We could not, of course, draw any conclusion from this if the concepts of pure mathematics which we find to be inherent in the structure of the universe were merely part of, or had been introduced through, the concepts of applied mathematics which we used to discover the workings of the universe. It would prove nothing if nature had merely been found to act in accordance with the concepts of applied mathematics; these concepts were specially and deliberately designed by man to fit the workings of nature. Thus it may still be objected that even our pure mathematics does not, in actual fact, represent a creation of our own minds so much as an effort, based on forgotten or subconscious memories, to understand the workings of nature. If so, it is not surprising that nature should be found to work according to the laws of pure mathematics. It cannot, of course, be denied that some of the concepts with which the pure mathematician works are taken direct from his experience of nature. An obvious instance is the concept of quantity, but this is so fundamental that it is hard to imagine any scheme of nature from which it was entirely excluded. Other concepts borrow at least something from experience; for instance, multi-dimensional geometry, which clearly originated out of experience of the three dimensions of space. If, however, the more intricate concepts of pure mathematics have been transplanted from the workings of nature, they must have been buried very deep indeed in our sub-conscious minds. This very controversial possibility is one which cannot be entirely dismissed, but it is exceedingly hard to believe that such intricate concepts as a finite curved space and an expanding space can have entered into pure mathematics through any sort of unconscious or sub-conscious experience of the workings of the actual universe. In any event, it can hardly be disputed that nature and our conscious mathematical minds work according to the same laws. She does not model her behaviour, so to speak, on that forced on us by our whims and passions, or on that of our muscles and joints, but on that of our thinking minds. This remains true whether our minds impress their laws on nature, or she impresses her laws on us, and provides a sufficient justification for thinking of the universe as being of mathematical design. Lapsing back again into the crudely anthropomorphic language we have already used, we may say that we have already considered with disfavour the possibility of the universe having been planned by a biologist or an engineer; from the intrinsic evidence of his creation, the Great Architect of the Universe now begins to appear as a pure mathematician.

Personally, I feel that this train of thought may, very tentatively, be carried a stage farther, although it is difficult to express it in exact words, again because our mundane vocabulary is circumscribed by our mundane experience. The terrestrial pure mathematician does not concern himself with material substance, but with pure thought. His creations are not only created by thought but consist of thought, just as the creations of the engineer consist of engines. And the concepts which now prove to be fundamental to our understanding of nature—a space which is finite; a space which is empty, so that one point differs from another solely in the properties of the space itself; four-dimensional, seven- and more dimensional spaces; a space which forever expands; a sequence of events which follows the laws of probability instead of the law of causation—or, alternately, a sequence of events which can only be fully and consistently described by going outside space and time—all these concepts seem to my mind to be structures of pure thought, incapable of realization in any sense which would properly be described as material.

For instance, anyone who has written or lectured on the finiteness of space is accustomed to the objection that the concept of a finite space is self-contradictory and nonsensical. If space is finite, our critics say, it must be possible to go out beyond this finite space, and what can we possibly find beyond it except more space, and so on *ad infinitum?*— which proves that space cannot be finite. And again, they say, if space is expanding, what can it possibly expand into, if not into more space?— which again proves that what is expanding can only be a part of space, so that the whole of space cannot expand.

The twentieth-century critics who make these comments are still in the state of mind of the nineteenth-century scientists; they take it for granted that the universe must admit of material representation. If we grant their premises, we must, I think, also grant their conclusion—that we are talking nonsense—for their logic is irrefutable. But modern science cannot possibly grant their conclusion; it insists on the finiteness of space at all costs. This, of course, means that we must deny the premises which our critics unknowingly assume. The universe cannot admit of material representation, and the reason, I think, is that it has become a mere mental concept.

It is the same, I think, with other more technical concepts, typified by the "exclusion principle," which seem to imply a sort of "action-at-a-distance" in both space and time—as though every bit of the universe knew what other distant bits were doing, and acted accordingly. To my mind, the laws which nature obeys are less suggestive of those which a machine obeys in its motion than of those which a musician obeys in writing a fugue, or a poet in composing a sonnet. The motions of electrons and atoms do not resemble those of the parts of a locomotive so much as those of the dancers in a cotillion. And if the "true essence of substances" is forever unknowable, it does not matter whether the cotillion is danced at a ball in real life, or on a cinematograph screen, or in a story of Boccaccio. If all this is so, then the universe can be best pictured, although still very imperfectly and inadequately, as consisting of pure thought, the thought of what, for want of a wider word, we must describe as a mathematical thinker.

And so we are led into the heart of the problem of the relation between mind and matter. Atomic disturbances in the distant sun cause it to emit light and heat. After "travelling through the ether" for eight minutes, some of this radiation may fall on our eyes, causing a disturbance on the retina, which travels along the optic nerve to the brain. Here it is perceived as a sensation by the mind; this sets our thoughts in action and results in, let us say, poetic thoughts about the sunset. There is a continuous chain, A, B, C, D . . . X, Y, Z, connecting A the poetic thought—through B the thinking mind, C the brain, D the optic nerve, and so on—with Z the atomic disturbance in the sun. The thought A results from the distant disturbance Z, just as the ringing of a bell results from pulling a distant bell-rope. We can understand how pulling a material rope can cause a material bell to ring because there is a material connection all the way. But it is far less easy to see how a disturbance of material atoms can cause a poetic thought to originate, because the two are so entirely dissimilar in nature.

For this reason, Descartes insisted that there could be no possible connection between mind and matter. He believed they were two entirely distinct kinds of entity, the essence of matter being extension in space, and that of mind being thought. And this led him to maintain that there were two distinct worlds, one of mind and one of matter, running, so to speak, independent courses on parallel rails without ever meeting.

Berkeley and the idealist philosophers agreed with Descartes that if mind and matter were fundamentally of different natures they could never interact. But they insisted that they continually do interact. Therefore, they argued, matter must be of the same nature as mind, so that, in the terminology of Descartes, the essence of matter must be thought rather than extension. Expressed in detail, their contention was that causes must be essentially of the same nature as their effects; if B on our chain produces A, then B must be of the same essential nature as A, and C as B, and so on. Thus Z also must be of the same essential nature as A. Now the only links of the chain of which we have any *direct* knowledge are our own thoughts and sensations A, B; we know of the existence and nature of the remote links X, Y, Z only by inference—from the effects they transmit to our minds through our senses. Berkeley, maintaining that the unknown distant links X, Y, Z, must be of the same nature as the known near links A, B, argued that they must be of the nature of thoughts or ideas, "since after all there is nothing like an idea except an idea." A thought or idea cannot, however, exist without a mind in which to exist. We may say an object exists in our minds while we are conscious of it, but this will not account for its existence during the time we are not conscious of it. The planet Pluto, for instance, was in existence long before any human mind suspected it, and was recording its existence on photographic plates long before any human eye saw it. Considerations such as these led Berkeley to postulate an Eternal Being, in whose mind all objects existed. And so, in the stately and sonorous diction of a bygone age, he summed up his philosophy in the words:

> All the choir of heaven and furniture of earth, in a word all those bodies which compose the mighty frame of the world, have not any substance without the mind. . . . so long as they are not actually perceived by me, or do not exist in my mind, or that of any other created spirit, they must either have no existence at all, or else subsist in the mind of some Eternal Spirit.

Modern science seems to me to lead, by a very different road, to a not altogether dissimilar conclusion. Biology, studying the connection between the earlier links of the chain, A, B, C, D, seems to be moving towards the conclusion that these are all of the same general nature. This is occasionally stated in the specific form that, as biologists believe C, D to be mechanical and material, A, B must also be mechanical and material, but apparently there would be at least equal warrant for stating it in the form that as A, B are mental, C, D must also be mental. Physical science, troubling little about C, D, proceeds directly to the far end of the chain; its business is to study the workings of X, Y, Z. And, as it seems to me, its conclusions suggest that the end links of the chain, whether we go to the cosmos as a whole or to the innermost structure of the atom, are of the same nature as A, B—of the nature of pure

16

A Universe of Pure Thought

T HIS MAY SUGGEST that we are proposing to discard realism entirely and enthrone a thoroughgoing idealism in its place. Yet this,
I think, would be too crude a statement of the situation. If it is true
that the "real essence of substances" is beyond our knowledge, then
the line of demarcation between realism and idealism becomes very
blurred indeed; it becomes little more than a relic of a past age in
which reality was believed to be identical with mechanism. Objective
realities exist because certain things affect your consciousness and
mine in the same way, but we are assuming something we have no
right to assume if we label them as either "real" or "ideal." The true
label is, I think, "mathematical," if we can agree that this is to connote
the whole of pure thought, and not merely the studies of the professional mathematician. Such a label does not imply anything as to what
things are in their ultimate essence, but merely something as to how
they behave.

The label we have selected does not, of course, relegate matter into
the category of hallucination or dreams. The material universe remains
as substantial as ever it was, and this statement must, I think, remain
true through all changes of scientific or philosophical thought.

For substantiality is a purely mental concept measuring the direct effect of objects on our sense of touch. We say that a stone or a motorcar
is substantial, while an echo or a rainbow is not. This is the ordinary
definition of the word, and it is a mere absurdity, a contradiction in
terms, to say that stones and motorcars can, in any way, become insubstantial, or even less substantial, because we now associate them with

mathematical formulae and thoughts, or kinks in empty space, rather than with crowds of hard particles. Dr Johnson is reported to have expressed his opinion on Berkeley's philosophy by dashing his foot against a stone and saying: "No, Sir, I disprove it thus." This little experiment had, of course, not the slightest bearing on the philosophical problem it claimed to solve; it merely verified the substantiality of matter. And, however science may progress, stones must always remain substantial bodies, just because they and their class form the standard by which we define the quality of substantiality.

It has been suggested that the lexicographer might really have disproved the Berkeleian philosophy if he had chanced to kick not a stone but a hat, in which some small boy had surreptitiously placed a brick; we are told that "the element of surprise is sufficient warrant for external reality," and that "a second warrant is permanence with change—permanence in your own memory, change in externality." This, of course, merely disproves the solipsist error of "all this is a creation of my own mind, and exists in no other mind," but it is hard to do anything in life which does not disprove this. The argument from surprise, and from new knowledge in general, is powerless against the concept of a universal mind of which your mind and mine, the mind which surprises and that which is surprised, are units or even excrescences. Each individual brain cell cannot be acquainted with all the thoughts which are passing through the brain as a whole.

Yet the fact that we possess no absolute extraneous standard against which to measure substantiality does not preclude our saying that two things have the same degree, or different degrees, of substantiality. If I dash my foot against a stone in my dreams, I shall probably waken up with a pain in my foot, to discover that the stone of my dreams was literally a creation of my mind and of mine alone, prompted by a nerve-impulse originating in my foot. This stone may typify the category of hallucinations or dreams; it is clearly less substantial than that which Johnson kicked. Creations of an individual mind may reasonably be called less substantial than creations of a universal mind. A similar distinction must be made between the space we see in a dream and the space of everyday life; the latter, which is the same for us all, is the space of the universal mind. It is the same with time, the time of waking life, which flows at the same even rate for us all, being the time of the universal mind. Again we may think of the laws to which phenomena conform in our waking hours, the laws of nature, as the laws of thought of a universal mind. The uniformity of nature proclaims the self-consistency of this mind.

This concept of the universe as a world of pure thought throws a new light on many of the situations we have encountered in our survey of modern physics. We can now see how the ether, in which all the events of the universe take place, could reduce to a mathematical abstraction and become as abstract and as mathematical as parallels of latitude and meridians of longitude. We can also see why energy, the fundamental entity of the universe, had again to be treated as a mathematical abstraction—the constant of integration of a differential equation.

The same concept implies, of course, that the final truth about a phenomenon resides in the mathematical description of it; so long as there is no imperfection in this, our knowledge of the phenomenon is complete. We go beyond the mathematical formula at our own risk; we may find a model or picture which helps us to understand it, but we have no right to expect this, and our failure to find such a model or picture need not indicate that either our reasoning or our knowledge is at fault. The making of models or pictures to explain mathematical formulae and the phenomena they describe is not a step towards, but a step away from reality; it is like making graven images of a spirit. And it is as unreasonable to expect these various models to be consistent with one another as it would be to expect all the statues of Hermes, representing the god in all his varied activities—as messenger, herald, musician, thief, and so on—to look alike. Some say that Hermes is the wind; if so, all his attributes are wrapped up in his mathematical description, which is neither more nor less than the equation of motion of a compressible fluid. The mathematician will know how to pick out the different aspects of this equation which represent the conveying and announcing of messages, the creation of musical tones, the blowing away of our papers, and so forth. He will hardly need statues of Hermes to remind him of them, although, if he is to rely on statues, nothing less than a whole row, all different, will suffice. All the same, some mathematical physicists are still busily at work, making graven images of the concepts of the wave-mechanics.

In brief, a mathematical formula can never tell us what a thing is, but only how it behaves; it can only specify an object through its properties. And these are unlikely to coincide *in toto* with the properties of any single macroscopic object of our everyday life.

This point of view brings us relief from many of the difficulties and apparent inconsistencies of present-day physics. We need no longer discuss whether light consists of particles or waves; we know all there is to be known about it if we have found a mathematical formula which accurately describes its behavior, and we can think of it as either particles or waves, according to our mood and the convenience of the moment. On our days of thinking of it as waves, we may, if we please, imagine an ether to transmit the waves, but this ether will vary from day to day; we have seen how it will vary each time our speed of motion varies. In the same way, we need not discuss whether the wave-system of a group of electrons exists in a three-dimensional space, or in a many-dimensional space, or not at all. It exists in a mathematical formula; this, and nothing else, expresses the ultimate reality, and we can picture it as representing waves in three, six, or more dimensions whenever we so please. We can also interpret it as not representing waves at all; in so doing, we shall be following Heisenberg and Dirac. It is generally simplest to interpret it as representing waves in a space having three dimensions for each electron, just as it is simplest to interpret the macroscopic universe as an array of objects in three dimensions only, and its phenomena as an array of events in four dimensions, but none of these interpretations possesses any unique or absolute validity.

If the universe is a universe of thought, then its creation must have been an act of thought. Indeed, the finiteness of time and space almost compel us, of themselves, to picture the creation as an act of thought; the determination of the constants such as the radius of the universe and the number of electrons it contained imply thought, whose richness is measured by the immensity of these quantities. Time and space, which form the setting for the thought, must have come into being as part of this act. Primitive cosmologies pictured a creator working in space and time, forging sun, moon, and stars out of already existent raw material. Modern scientific theory compels us to think of the creator as working outside time and space—which are part of his creation—just as the artist is outside his canvas. It accords with the conjecture of Augustine: *"Non in tempore, sed cum tempore, finxit Deus mundum."* Indeed, the doctrine dates back as far as Plato:

> Time and the heavens came into being at the same instant, in order that, if they were ever to dissolve, they might be dissolved together. Such was the mind and thought of God in the creation of time.

And yet, so little do we understand time that perhaps we ought to compare the whole of time to the act of creation, the materialisation of the thought.

Today there is a wide measure of agreement which, on the physical side of science, approaches almost to unanimity that the stream of knowledge is heading towards a nonmechanical reality; the universe begins to look more like a great thought than like a great machine. Mind no longer appears as an accidental intruder into the realm of matter; we are beginning to suspect that we ought rather to hail it as the creator and governor of the realm of matter—not, of course, our individual minds, but the mind in which the atoms out of which our individual minds have grown exist as thoughts.

The new knowledge compels us to revise our hasty first impressions that we had stumbled into a universe which either did not concern itself with life or was actively hostile to life. The old dualism of mind and matter, which was mainly responsible for the supposed hostility, seems likely to disappear, not through matter becoming in any way more shadowy or insubstantial than heretofore, or through mind becoming resolved into a function of the working of matter, but through substantial matter resolving itself into a creation and manifestation of mind. We discover that the universe shows evidence of a designing or controlling power that has something in common with our own individual minds—not, so far as we have discovered, emotion, morality, or aesthetic appreciation, but the tendency to think in the way which, for want of a better word, we describe as mathematical. And while much in it may be hostile to the material appendages of life, much also is akin to the fundamental activities of life; we are not so much strangers or intruders in the universe as we at first thought. Those inert atoms in the primeval slime which first began to foreshadow the attributes of life were putting themselves more, and not less, in accord with the fundamental nature of the universe.

So at least we are tempted to conjecture today, and yet who knows how many more times the stream of knowledge may turn on itself? And with this reflection before us, we may well conclude by adding what might well have been interlined into every paragraph: that everything that has been said, and every conclusion that has been tentatively put forward, is quite frankly speculative and uncertain. We have tried to discuss whether present-day science has anything to say on certain difficult questions, which are perhaps set forever beyond the reach of human understanding. We cannot claim to have discerned more than a very faint glimmer of light at the best; perhaps it was wholly illusory, for certainly we had to strain our eyes very hard to see anything at all. So that our main contention can hardly be that the science of today has a pronouncement to make, perhaps it ought rather to be that science should leave off making pronouncements: the river of knowledge has too often turned back on itself.

Editor's Footnote:

We don't have to agree with everything Jeans said in order to point out that the idea of the physical realm being a "materialization of thought" has extremely wide support from the perennial philosophy. As Huston Smith points out in *Forgotten Truth,* the perennial philosophy has always maintained that matter is a crystallization or a precipitation of mind (ontologically, not chronologically). Actually, this "precipitation" process runs throughout the Great Chain of Being. With reference to the diagram in Chapter One, we can explain it like this: If you start with the spiritual realm (Level 5) and *subtract* "E," you get soul; if you then *subtract* "D," you get mind; subtract "C" and you get life; subtract "B" and you get matter. This *subtraction process* is a progressive precipitation of the lower from the higher, a process called "involution"; each junior dimension is thus a *reduced subset* of its senior dimension. The reverse of this subtraction, precipitation, or involution process is simply *evolution,* or the unfolding of successively senior dimensions from their prior or involutionary enfoldment in the lower domains (where they exist, as Aristotle would have it, *in potentia,* although nothing in the lower gives any evidence that a higher can break through it and emerge transcendentally beyond its domains). This is why evolution, *vis à vis* the lower, is an *addition* or *creative emergence* of successively higher domains from (or rather through) the junior dimensions. Involution, we may speculate, gave rise to the "Big Bang," where the material realm blew into existence via a concrete precipitation of the higher (although at this point still ontologically implicit) realms, and the universe has been evolving back or upwards ever since, producing thus far matter, then life, then mind (and in some saints and sages, a *conscious* realization or concrescence of soul and then spirit).

The significant point: every physicist in this volume was profoundly struck by the fact that the natural realm (Levels 1 and 2) obeys in some sense the laws or forms of mathematics, or, in general, obeys some sort of archetypal mental-forms (which reside at Levels 3 and 4). But that is exactly what would be expected if the natural realms are a reduced subset or precipitate of the mind-soul realms; the child obeys its ontic parents. Heisenberg and Pauli looking for the archetypal forms which underlie the material realm; de Broglie claiming mind-forms had to precede (ontically) matter forms; Einstein and Jeans finding a central mathematical form to the cosmos—all of that becomes perfectly understandable in this light.

Because the natural realms are a *reduced* subset of, or are ontically *less than,* the mental-soul realms, then all fundamental natural processes can be essentially represented mathematically, but not all mathematical forms have a material application. That is, of the almost infinite number of mathematical schemes existing implicitly in the mental-soul realms, only a rather small, finite number actually crystallize or precipitate in and as the material realm. Put another way, because the material realms are ontically much less than the mental, only the relatively *simpler* mental-soul forms show up in, or precipitate as, the material realm. And this leads exactly to *the* guiding principle that every one of these physicists followed in trying to discover the mental laws governing material phenomena: of all possible mathematical schemes that might explain physical data, *choose the simplest and most elegant.* Einstein put it perfectly: "Nature is the *realization* [crystallization or precipitation] of the simplest conceivable mathematical ideas." This does not mean that matter *is* an idea, pure and simple; it means that whatever Matter is, is a reduced, subtracted, or condensed version of whatever Idea is. Matter is a Platonic shadow, if you wish, but a shadow that, as Jeans says, bears *some* of the forms of the ontically higher domains, in this case, mathematical forms.

Finally, this explains why all these physicists maintained that mathematical laws cannot be deduced from mere sensory-physical-empirical data: you cannot deduce or derive the higher from the lower. To *check* whether a particular mathematical scheme correctly applies to some physical realm, we must use the physical senses (or their instrumental extensions); to *find* that mathematical scheme in the first place, however, we use mind and only mind. What we are doing (using the eye of reason) is searching through the mental universe to see which schemes or forms might have crystallized in and as this particular physical universe (which we then check with the eye of flesh). Thus the criteria for establishing the truth of a physical theory: *vis à vis* mind, it must be *coherent* (free of self-contradiction); *vis à vis* physical data, it must *correspond* (match or fit evidence); if two theories equally meet those criteria (which happens very often), then choose the simpler and more elegant. The empiricists want only correspondence theories of truth; the idealists, only coherence theories; whereas both are equally important, and simple elegance or beauty the final crown. I think this is why Heisenberg so often quoted "The simple is the seal of the true" and "Beauty is the splendor of the truth."

—KW

PLANCK

MAX PLANCK
(1858–1947)

IT WAS MAX PLANCK'S bold, brilliant, daring, and wholly unprecedented leap of genius that, in 1900, ushered in the entire quantum revolution, for it was Planck who hit upon the idea that nature is not continuous, but rather comes in discrete packets or quanta. Justly regarded as the father of modern quantum theory, Planck was awarded the Nobel Prize in Physics in 1918.

Of Planck, who was deeply respected and loved by all his colleagues, Albert Einstein had these memorable words: "The longing to behold harmony is the source of the inexhaustible patience and perseverance with which Planck has devoted himself to the most general problems of our science, refusing to let himself be diverted to more grateful and more easily attained ends. I have often heard colleagues try to attribute this attitude of his to extraordinary will-power and discipline—wrongly, in my opinion. The state of mind which enables a man to do work of this kind is akin to that of the religious worshipper or the lover; the daily effort comes from no deliberate intention or program, but straight from the heart. There he sits, our beloved Planck, and smiles inside himself at my childish playing-about with the lantern of Diogenes. Our affection for him needs no thread-bare explanation. May the love of science continue to illumine his path in the future and lead him to the solution of the most important problems in present-day physics, which he has himself posed and done so much to solve."

The following sections are taken from *Where Is Science Going?* (New York: Norton, 1932).

17

The Mystery of Our Being

W E MIGHT NATURALLY ASSUME that one of the achievements of science would have been to restrict belief in miracle. But it does not seem to do so. The tendency to believe in the power of mysterious agencies is an outstanding characteristic of our own day. This is shown in the popularity of occultism and spiritualism and their innumerable variants. Though the extraordinary results of science are so obvious that they cannot escape the notice of even the most unobservant man in the street, yet educated as well as uneducated people often turn to the dim region of mystery for light on the ordinary problems of life. One would imagine that they would turn to science, and it is probably true that those who do so are more intensely interested in science and are perhaps greater in number than any corresponding group of people in former times; still the fact remains that the drawing power of systems which are based on the irrational is at least as strong and as widespread as ever before, if not more so.

How is this peculiar fact to be explained? Is there, in the last analysis, some basically sound foothold for this belief in miracle, no matter how bizarre and illogical may be the outer forms it takes? Is there something in the nature of man, some inner realm, that science cannot touch? Is it so that when we approach the inner springs of human action science cannot have the last word? Or, to speak more concretely, is there a point at which the causal line of thought ceases and beyond which science cannot go?

This brings us to the kernel of the problem in regard to free will. And I think that the answer will be found automatically suggested by the questions which I have just asked.

The fact is that there is a point, one single point in the immeasurable world of mind and matter, where science and therefore every causal method of research is inapplicable, not only on practical grounds but also on logical grounds, and will always remain inapplicable. This point is the individual ego. [In the German philosophic tradition in which Planck is writing, the term "ego" means "the I," or the inward sense of "I-ness" constituting your sense of self. It doesn't mean "egotistical," but rather that irreducible, immediate, inward sense of consciousness or awareness.—Ed. Note] It is a small point in the universal realm of being, but, in itself, it is a whole world, embracing our emotional life, our will, and our thought. This realm of the ego is, at once, the source of our deepest suffering and, at the same time, of our highest happiness. Over this realm, no outer power of fate can ever have sway, and we lay aside our own control and responsibility over ourselves only with the laying aside of life itself.

Here is the place where the freedom of the will comes in and establishes itself, without usurping the right of any rival. Being emancipated thus, we are at liberty to construct any miraculous background that we like in the mysterious realm of our own inner being, even though we may be at the same time the strictest scientists in the world, and the strictest upholders of the principle of causal determinism. It is from this autarchy of the ego that the belief in miracles arises, and it is to this source that we are to attribute the widespread belief in irrational explanations of life. The existence of that belief in the face of scientific advance is a proof of the inviolability of the ego by the law of causation in the sense which I have mentioned. I might put the matter in another way and say that the freedom of the ego here and now, and its independence of the causal chain, is a truth that comes from the immediate dictate of the human consciousness.

And what holds good for the present moment of our being holds good also for our own future conduct in which the influences of our present ego plays a part. The road to the future always starts in the present. It is, here and now, part and parcel of the ego. And for that reason, the individual can never consider his own future purely and exclusively from the causal standpoint. That is the reason why fancy plays such a part in the construction of the future. It is in actual recognition of this profound fact that people have recourse to the palmist and the clairvoyant to satisfy their individual curiosity about their own future. It is also on this fact that dreams and ideals are based, and here the human being finds one of the richest sources of inspiration.

Science thus brings us to the threshold of the ego and there leaves us to ourselves. Here it resigns us to the care of other hands. In the conduct of our own lives, the causal principle is of little help; for by the iron law of logical consistency, we are excluded from laying the causal foundations of our own future or foreseeing that future as definitely resulting from the present.

But mankind has need of fundamental postulates for the conduct of everyday existence, and this need is far more pressing than the hunger for scientific knowledge. A single deed often has far more significance for a human being than all the wisdom of the world put together. And, therefore, there must be another source of guidance than mere intellectual equipment. The law of causation is the guiding rule of science, but the Categorical Imperative—that is to say, the dictate of duty—is the guiding rule of life. Here intelligence has to give place to character, and scientific knowledge to religious belief. And when I say religious belief here I mean the word in its fundamental sense. And the mention of it brings us to that much discussed question of the relation between science and religion. It is not my place here nor within my competency to deal with that question. Religion belongs to that realm that is inviolable before the law of causation and, therefore, closed to science. The scientist as such must recognize the value of religion as such, no matter what may be its forms, so long as it does not make the mistake of opposing its own dogmas to the fundamental law upon which scientific research is based, namely, the sequence of cause and effect in all external phenomena. In conjunction with the question of the relations between religion and science, I might also say that those forms of religion which have a nihilist attitude to life are out of harmony with the scientific outlook and contradictory to its principles. All denial of life's value for itself and for its own sake is a denial of the world of human thought and, therefore, in the last analysis, a denial of the true foundation not only of science but also of religion. I think that most scientists would agree to this and would raise their hands against religious nihilism as destructive of science itself.

There can never be any real opposition between religion and science. Every serious and reflective person realizes, I think, that the religious element in his nature must be recognized and cultivated if all the powers of the human soul are to act together in perfect balance and harmony. And, indeed, it was not by any accident that the greatest thinkers of all ages were also deeply religious souls, even though they made no public show of their religious feeling. It is from the cooperation of the understanding with the will that the finest fruit of philosophy has arisen, namely, the ethical fruit. Science enhances the moral values of life because it furthers a love of truth and reverence—love of truth displaying itself in the constant endeavor to arrive at a more exact knowledge of the world of mind and matter around us, and reverence, because every advance in knowledge brings us face to face with the mystery of our own being.

"THE PURE RATIONALIST HAS NO PLACE HERE"

Planck: The churches appear to be unable to supply that spiritual anchorage which so many people are seeking. And so the people turn in other directions. The difficulty which organized religion finds in appealing to the people nowadays is that its appeal necessarily demands the believing spirit, or what is generally called Faith. In an all-round state of skepticism this appeal receives only a poor response. Hence you have a number of prophets offering substitute wares.

Murphy: Do you think that science in this particular might be a substitute for religion?

Planck: Not to a skeptical state of mind; for science demands also the believing spirit. Anybody who has been seriously engaged in scientific work of any kind realizes that over the entrance to the gates of the temple of science are written the words: *Ye must have faith.* It is a quality which the scientists cannot dispense with.

The man who handles a bulk of results obtained from an experimental process must have an imaginative picture of the law that he is pursuing. He must embody this in an imaginary hypothesis. The reasoning faculties alone will not help him forward a step, for no order can emerge from that chaos of elements unless there is the constructive quality of mind which builds up the order by a process of elimination and choice. Again and again the imaginary plan on which one attempts to build up that order breaks down and then we must try another. This imaginative vision and faith in the ultimate success are indispensable. The pure rationalist has no place here.

Murphy: How far has this been verified in the lives of great scientists? Take the case of Kepler, whose 300th anniversary we were celebrating, you remember, that evening when Einstein gave his lecture at the Academy of Science. Wasn't there something about Kepler having made certain discoveries, not because he set out after them with his constructive imagination, but rather because he was concerned about the dimensions of wine barrels and was wondering which shapes would be the most economic containers?

Planck: These stories circulate in regard to nearly everybody whose name is before the public. As a matter of fact, Kepler is a magnificent example of what I have been saying. He was always hard up. He had to suffer disillusion after disillusion and even had to beg for the payment of the arrears of his salary by the Reichstag in Regensburg. He had to undergo the agony of having to defend his own mother against a public indictment of witchcraft. But one can realize, in studying his life, that what rendered him so energetic and tireless and productive was the profound faith he had in his own science, not the belief that he could eventually arrive at an arithmetical synthesis of his astronomical observations, but rather the profound faith in the existence of a definite plan behind the whole of creation. It was because he believed in that plan that his labor was felt by him to be worthwhile and also in this way, by never allowing his faith to flag, his work enlivened and enlightened his dreary life. Compare him with Tycho de Brahe. Brahe had the same material under his hands as Kepler, and even better opportunities, but he remained only a researcher, because he did not have the same faith in the existence of the eternal laws of creation. Brahe remained only a researcher; but Kepler was the creator of the new astronomy.

Another name that occurs to me in this connection is that of Julius Robert Mayer. His discoveries were hardly noticed, because in the middle of last century there was a great deal of skepticism, even among educated people, about the theories of natural philosophy. Mayer kept on and on, not because of what he had discovered and could prove, but because of what he believed. It was only in 1869 that the Society of German Physicists and Physicians, with Helmholtz at their head, recognized Mayer's work.

Murphy: You have often said that the progress of science consists in the discovery of a new mystery the moment one thinks that something fundamental has been solved.

Planck: This is undoubtedly true. Science cannot solve the ultimate mystery of nature. And that is because, in the last analysis, we ourselves are part of nature and, therefore, part of the mystery that we are trying to solve. Music and art are, to an extent, also attempts to solve or at least to express the mystery. But to my mind, the more we progress with either, the more we are brought into harmony with all nature itself. And that is one of the great services of science to be individual.

Murphy: Goethe once said that the highest achievement to which the human mind can attain is an attitude of wonder before the elemental phenomena of nature.

Planck: Yes, we are always being brought face to face with the irrational. Else we couldn't have faith. And if we did not have faith but could solve every puzzle in life by an application of the human reason, what an unbearable burden life would be. We should have no art and no music and no wonderment. And we should have no science; not only because science would thereby lose its chief attraction for its own followers—namely, the pursuit of the unknowable—but also because science would lose the cornerstone of its own structure, which is the direct perception by consciousness of the existence of external reality. As Einstein has said, you could not be a scientist if you did not know that the external world existed in reality, but that knowledge is not gained by any process of reasoning. It is a direct perception and, therefore, in its nature akin to what we call Faith. It is a metaphysical belief. Now that is something which the skeptic questions in regard to religion, but it is the same in regard to science. However, there is this to be said in favor of theoretical physics, that it is a very active science and does make an appeal to the lay imagination. In that way it may, to some extent, satisfy the metaphysical hunger which religion does not seem capable of satisfying nowadays. But this would be entirely by stimulating the religious reaction indirectly. Science as such can never really take the place of religion.

PAULI

WOLFGANG PAULI
(1900–1958)

IN TERMS OF SHEER intellectual brilliance, Wolfgang Pauli was probably second to no physicist of this or any period (according to Max Born, Pauli's genius exceeded even that of Einstein). Intellectual sloppiness or logical inconsistency would bring down the wrath of Pauli on the poor soul unfortunate enough to be its author. He was a brilliant and ruthless critic of ideas, and virtually every physicist of his generation looked to the mind of Wolfgang Pauli as one of *the* mandatory tests to pass if a theory had any chance of survival. Pauli's own positive contributions were profound and numerous, including the famous "exclusion principle" and the prediction of the existence of the neutrino some two decades before it was discovered. He received the Nobel Prize in Physics in 1945.

In spite of, or rather precisely because of, Pauli's analytical and intellectual brilliance, he insisted that rationality had to be supplemented with the mystical. I had originally planned to include in this section Pauli's essay, "The Influence of Archetypal Ideas on Kepler's Construction of Scientific Theories," which sets forth his Platonic-Pythagorean worldview, and which was written in collaboration with C.G. Jung. But his lifetime friend and colleague, Werner Heisenberg, wrote a beautiful summary of Pauli's position, which is not only briefer but considerably more elegant reading, and so I have presented that instead ("Wolfgang Pauli's Philosophical Outlook," chapter 3 in *Across the Frontiers*).

18

Embracing the Rational
and the Mystical

WOLFGANG PAULI'S WORK in theoretical physics allows us only
at a few places to recognize the philosophical background from
which it has arisen. To his colleagues, Pauli appears preeminently the
brilliant physicist, always inclined to the most incisive formulations,
who, by significant new ideas, by an analysis of existing findings clear
down to the last detail, and by unsparing criticism of every obscurity
and inexactitude in proposed theories, has decisively influenced and en-
riched the physics of the present century. If we wanted to construct a
basic philosophical attitude from these scientific utterances of Pauli's, at
first we would be inclined to infer from them an extreme rationalism
and a fundamentally skeptical point of view. In reality, however, behind
this outward display of criticism and skepticism lay concealed a deep
philosophical interest even in those dark areas of reality or the human
soul which elude the grasp of reason. And while the power of fascination
emanating from Pauli's analyses of physical problems was admittedly
due in some measure to the detailed and penetrating clarity of his formulations, the rest was derived from a constant contact with the field of
creative spiritual processes, for which no rational formulation as yet
exists. Very early in his career, Pauli had followed the road of skepticism
based on rationalism right to the end, to a skepticism about skepticism,
and he then tried to trace out those elements of the cognitive process
that precede a rational understanding in depth. There are two essays in
particular from which the essentials of Pauli's philosophical attitude
may be gathered: an article on "The Influence of Archetypal Ideas on

Kepler's Construction of Scientific Theories" and a lecture on "Science and Western Thought." From these two sources and from his letters and other pronouncements, we shall try to obtain a picture of Pauli's philosophical point of view.

A first central topic of philosophical reflection for Pauli was the process of knowledge itself, especially that of natural knowledge, which ultimately finds its rational expression in the establishment of mathematically formulated laws of nature. Pauli was not satisfied with the purely empiricist view whereby natural laws can be drawn solely from the data of experience. He allied himself, rather, with those who "emphasize the role of intuition and the direction of attention in framing the concepts and ideas necessary for the establishing of a system of natural laws (i.e., a scientific theory)—ideas which in general go far beyond mere experience." He therefore sought for a connecting link between sense perceptions on the one hand and concepts on the other:

> All consistent thinkers have come to the conclusion that pure logic is fundamentally incapable of constructing such a linkage. The most satisfactory course, it seems, is to introduce at this point the postulate of an order of the cosmos distinct from the world of appearances, and not a matter of our choice. Whether we speak of natural objects participating in the Ideas or of the behavior of metaphysical, i.e., intrinsically real things, the relation between sense perception and Idea remains a consequence of the fact that both the soul and what is known in perception are subject to an order objectively conceived.

The bridge leading from the initially unordered data of experience to the Ideas is seen by Pauli in certain primeval images preexisting in the soul, the archetypes discussed by Kepler and also by modern psychology. These primeval images—here Pauli is largely in agreement with the views of Jung—should not be located in consciousness or related to specific rationally formulable ideas. It is a question, rather, of forms belonging to the unconscious region of the human soul, images of powerful emotional content, which are not thought but are beheld, as it were, pictorially. The delight one feels on becoming aware of a new piece of knowledge arises from the way such preexisting images fall into congruence with the behavior of external objects.

This view of natural knowledge is notoriously derived in its essentials from Plato, and it penetrated into Christian thought by way of neo-

Platonism (Plotinus, Proclus). Pauli seeks to clarify it by pointing out that even Kepler's conversion to the Copernican theory, which marks the beginning of modern natural science, was decisively affected by certain primeval images or archetypes. He cites this passage from Kepler's *Mysterium Cosmographicum:* "The image of the triune God is in the sphere, namely of the Father in the center, of the Son in the outer surface and of the Holy Ghost in the uniformity of connection between point and intervening space or surroundings." The motion directed from the center to the outer surface is, for Kepler, the emblem of creation. This symbol, most intimately associated with the Holy Trinity and described by Jung as a *mandala*, finds an imperfect realization, for Kepler, in the physical world: the sun in the center of the system of planets, surrounded by the heavenly bodies (which Kepler still thought to be animate). Pauli believes that to Kepler, the persuasiveness of the Copernican system is due primarily to its correspondence with the symbol described and only secondarily to the data of experience.

Pauli considers, moreover, that Kepler's symbol illustrates quite generally the attitude from which contemporary science has arisen. "From an inner center, the mind seems to move outward in a sort of extraversion into the physical world, in which all happenings are assumed to be automatic, so that the spirit serenely encompasses this physical world, as it were, with its Ideas." Thus the natural science of the modern era involves a Christian elaboration of the "lucid mysticism" of Plato, in which the unitary ground of spirit and matter is sought in the primeval images, and in which understanding has found its place in its various degrees and kinds, even to knowledge of the word of God. But Pauli adds a warning: "This mysticism is so lucid that it sees out beyond many obscurities, which we moderns dare not and cannot do."

He therefore contrasts the outlook of Kepler with that of a contemporary, the English physician Robert Fludd, with whom Kepler had entered into a violent polemic about the application of mathematics to experience refined by quantitative measurement. Fludd is here the exponent of an archaically magic description of nature, of the kind practiced by medieval alchemy and the secret societies that arose from it.

The elaboration of Plato's thought had led, in neo-Platonism and Christianity, to a position where matter was characterized as void of Ideas. Hence, since the intelligible was identical with the good, matter was identified with evil. But in the new science the world-soul was finally replaced by the abstract mathematical law of nature. Against this one-sidedly spiritualizing tendency the alchemistical philosophy, championed here by Fludd, represents a certain counterpoise. In the alchemistic view "there dwells in matter a spirit awaiting release. The alchemist in his laboratory is constantly involved in nature's course, in such wise that the real or supposed chemical reactions in the retort are mystically identified with the psychic processes in himself, and are called by the same names. The release of the substance by the man who transmutes it, which culminates in the production of the philosopher's stone, is seen by the alchemist, in light of the mystical correspondence of macrocosmos and microcosmos, as identical with the saving transformation of the man by the work, which succeeds only 'Deo concedente.' " The governing symbol for this magical view of nature is the quaternary number, the so-called "tetractys" of the Pythagoreans, which is put together out of two polarities. The division is correlated with the dark side of the world (matter, the Devil), and the magical view of nature also embraces this dark region.

Neither of these two lines of development, taking their rise from Plato and Christian philosophy on the one hand and from medieval alchemy on the other, could later escape disintegration into opposing systems of thought. Platonic thought, originally directed toward the unity of matter and spirit, leads eventually to a cleavage into the scientific and the religious views of the world, while the spiritual trend determined by gnosticism and alchemy produces scientific chemistry on the one hand and, on the other, a religious mysticism again divorced from material processes, as in Jakob Böhme.

In these mutually divergent and yet cognate lines of spiritual development, Pauli discerns complementary relationships which have determined Western thought from the outset and which today, now that the logical possibility of such relations has become fathomable to us through quantum mechanics, are more easily intelligible to us than they were to earlier ages. In scientific thinking, which is especially characteristic of the West, the soul turns outward and asks after the why of things. "Why is the one reflected in the many, what is the reflector and what the reflected, why did not the one remain alone?" Mysticism, conversely, which is equally at home in both East and West, endeavors to experience the unity of things, in that it seeks to penetrate beyond multiplicity, which it treats as an illusion. The scientific pursuit of knowledge led in the nineteenth century to the limiting concept of an objective material world, independent of all observation, while at the end point of the mystical experience there stands as a limiting situation the soul entirely divorced from all objects and united with the divine. Pauli sees Western thought as strung out, so to speak, between these two limiting ideas. "There will always be two attitudes dwelling in the soul of man, and the one will always carry the other already within it, as the seed of its opposite. Hence arises a sort of dialectical process, of which we know not wither it leads us. I believe that as Westerners we must entrust ourselves to this process, and acknowledge the two opposites to be complementary. In allowing the tension of the opposites to persist, we must also recognize that in every endeavor to know or solve we depend upon factors which are outside our control, and which religious language has always entitled 'grace.' "

When, in the spring of 1927, opinions on the interpretation of quantum mechanics were taking on rational shape and Bohr was forging the concept of complementarity, Pauli was one of the first physicists to decide unreservedly for the new possibility of interpretation. The characteristic feature of this interpretation—namely, that in every experiment, every incursion into nature, we have the choice of which aspect of nature we want to make visible, but that we simultaneously make a sacrifice, in that we must forgo other such aspects—this coupling of "choice and sacrifice," proved spontaneously congenial to Pauli's philosophical outlook. In the center of his philosophical thinking here there was always the wish for a unitary understanding of the world, a unity incorporating the tension of opposites, and he hailed the interpretation of quantum theory as a new way of thinking, in which the unity can perhaps be more easily expressed than before. In the alchemistic philosophy, he had been captivated by the attempt to speak of material and psychical processes in the same language. Pauli came to think that in the abstract territory traversed by modern atomic physics and modern psychology, such a language could once more be attempted:

> For I suspect that the alchemistical attempt at a unitary psycho-physical language miscarried only because it was related to a visible concrete reality. But in physics today we have an invisible reality (of atomic objects) in which the observer intervenes with a certain freedom (and is thereby confronted with the alternatives of "choice and sacrifice"); in the psychology of the unconscious we have processes which cannot always be unambiguously ascribed to a particular subject. The attempt at a psychophysical monism seems to me now essentially more promising, given that the relevant unitary language (unknown as yet, and neutral in

regard to the psychophysical antithesis) would relate to a deeper invisible reality. We should then have found a mode of expression for the unity of all being, transcending the causality of classical physics as a form of correspondence (Bohr); a unity of which the psychophysical interrelation, and the coincidence of *a priori* instinctive forms of ideation with external perceptions, are special cases. On such a view, traditional ontology and metaphysics become the sacrifice, but the choice falls on the unity of being.

Among the studies to which Pauli was prompted by the philosophical labors just referred to, it was those on the symbolism of the alchemists which left particularly lasting traces behind, as can be seen on occasion from utterances in his letters. In the theory of elementary particles, for instance, he is delighted with the various intertwined fourfold symmetries, which he at once relates to the tetractys of the Pythagoreans; or he writes: "Bisection and lessening of symmetry, that's the poodle's core. Dividing in two is a very old attribute of the devil (the word 'doubtful' is supposed to have originally meant 'twofold')." The philosophical systems from the period after the Cartesian bifurcation were less congenial to him. The Kantian employment of the *a priori* concept he criticizes in very decided terms, since Kant had used this expression for rationally fixable forms of intuition or forms of thought. He expressly warns that "one should never declare theses laid down by rational formulation to be the only possible presuppositions of human reason." Pauli, on the contrary, brings the *a priori* elements of natural science into intimate connection with the primeval images, the archetypes of Jungian psychology, which do not necessarily have to be regarded as innate but may be slowly variable and relative to a given cognitive situation. On this point, therefore, the views of Pauli and Jung diverge from those of Plato, who looked on the primary images as existing unchangeably and independent of the human soul. But, in each case, these archetypes are consequences or evidences of a general order of the cosmos, embracing matter and spirit alike.

In regard to this unitary order of the cosmos, which still cannot be rationally formulated, Pauli is also skeptical of the Darwinian opinion, extremely widespread in modern biology, whereby the evolution of species on earth is supposed to have come about solely according to the laws of physics and chemistry, through chance mutations and their subsequent effects. He feels this scheme to be too narrow and considers the possibility of more general connections, which can neither be fitted into the general conceptual scheme of causal structures nor be properly described by the term "chance." Repeatedly, we encounter in Pauli an endeavor to break out of the accustomed grooves of thought in order to come closer, by new paths, to an understanding of the unitary structure of the world.

It goes without saying that Pauli, in his wrestlings with the "One," was also continually obliged to come to terms with the concept of God; when he writes in a letter of the "theologians, to whom I stand in the archetypal relation of a hostile brother," this remark is certainly also seriously intended. Little as he was in the position of simply living and thinking within the tradition of one of the old religions, so equally little was he prepared to go over to a naïve, rationalistically grounded atheism. No better account could well be given of Pauli's attitude to this most general of questions than that which he himself has offered in the concluding section of his lecture on science and Western thought:

> I believe, however, that to anyone for whom a narrow rationalism has lost its persuasiveness, and to whom the charm of a mystical attitude, experiencing the outer world in its oppressive multiplicity as illusory, is also not powerful enough, nothing else remains but to expose oneself in one way or another to these intensified oppositions and their conflicts. Precisely by doing so, the inquirer can also more or less consciously tread an inner path to salvation. Slowly there then emerge internal images, fantasies or Ideas to compensate the outer situation, and which show an approach to the poles of the antitheses to be possible. Warned by the miscarriage of all premature endeavors after unity in the history of human thought, I shall not venture to make predictions about the future. But, contrary to the strict division of the activity of the human spirit into separate departments—a division prevailing since the nineteenth century—I consider the ambition of overcoming opposites, including also a synthesis embracing both rational understanding and the mystical experience of unity, to be the mythos, spoken or unspoken, of our present day and age.

EDDINGTON

Sir Arthur Eddington

(1882–1944)

S IR ARTHUR EDDINGTON made important contributions to the theoretical physics of the motion, evolution, and internal constitution of
stellar systems. He was one of the first theorists to grasp fully relativity
theory, of which he became a leading exponent. No mere armchair theorist, Eddington led the famous expedition that photographed the solar
eclipse which offered the first proof of Einstein's relativity theory. For
his outstanding contributions, he was knighted in 1930.

The following sections are taken from *Science and the Unseen World*
(New York: Macmillan, 1929), *New Pathways in Science* (New York:
Macmillan, 1935), and *The Nature of the Physical World* (New York:
Macmillan, 1929). Of all the physicists in this volume, Eddington was
probably the most eloquent writer; with Heisenberg, the most accomplished philosopher; and with Schroedinger, the most penetrating mystic. Moreover, he possessed an exquisite intellectual wit, evidenced on
almost every page of his writings (it sometimes takes the reader a while
to realize just how humorous Eddington is being, so set your mind in
that direction now). I have divided his topics into three rough sections,
the first dealing with the shadowy limitations of physical science, the
second with the necessity to equate the reality behind the shadows with
consciousness itself, and the third, his famous defense of mysticism.

19

Beyond the Veil of Physics

[B EFORE WE ENTER into Eddington's sophisticated arguments, it is necessary to allow him to speak for himself as to what exactly he is, and especially is *not,* trying to accomplish. His masterpiece, *The Nature of the Physical World,* was so persuasive and eloquent on the themes of physics and mysticism that his actual conclusion—namely, that the two are dealing with entirely different issues and domains—was quickly overlooked by the public (and especially the theologians), and Eddington earned the wholly undeserved reputation of claiming that the new physics supported (or even offered proof for) a mystical worldview. This rankled Eddington no end, for it was exactly the opposite of his views. When Bertrand Russell unleashed his considerable philosophic wit on Eddington's supposed derivation of mysticism from physics, Sir Arthur could no longer contain himself, and, in *New Pathways in Science,* Eddington answered sharply:]

My last round will be with Bertrand Russell. I think that he, more than any other writer, has influenced the development of my philosophical views, and my debt to him is great indeed. But this is necessarily a quarrelsome chapter, and I must protest against the following accusation: Sir Arthur Eddington deduces religion from the fact that atoms do not obey the laws of mathematics. Sir James Jeans deduces it from the fact that they do.

Russell here attributes to me a view of the basis of religion which I have strongly opposed whenever I have touched on the subject. I gather from what precedes this passage that Russell is really referring to my views on free will, which he appears to regard as equivalent to religion; even so, the statement is far from true. I have not suggested that either religion or free will can be deduced from modern physics; I have limited myself to showing that certain difficulties in reconciling them with physics have been removed. If I found a prevailing opinion that Russell could not be a competent mathematician because he had claimed to square the circle, I might, in defending him, point out that the report that he had made such a claim was without foundation. Would it be fair to say that I deduce that Russell is a competent mathematician from the fact that he has not claimed to square the circle?

One might have regarded the foregoing as a casual sacrifice of accuracy to epigram, but other passages make the same kind of accusation:

> It will be seen that Eddington, in this passage, does not infer a definite act of creation by a Creator. His only reason for not doing so is that he does not like the idea. The scientific argument leading to the conclusion which he rejects is much stronger than the argument in favour of free will, since that is based on ignorance, whereas the one we are now considering is based upon knowledge. This illustrates the fact that the theological conclusions drawn by scientists from their science are only such as please them, and not such as their appetite for orthodoxy is insufficient to swallow, although the argument would warrant them.

Memories are short, and one man is sometimes saddled with another man's opinions. It seems worthwhile, therefore, to give quotations showing how completely Russell has misstated my view of the relation of science and religion. I think that every book or article in which I have touched on religion is represented in these extracts, except an early essay which does not provide a passage compact enough to quote.

> The starting-point of belief in mystical religion is a conviction of significance or, as I have called it earlier, the sanction of a striving in the consciousness. This must be emphasised because appeal to intuitive conviction of this kind has been the foundation of religion through all ages and I do not wish to give the impression that we have now found something new and more scientific to substitute. I repudiate the idea of proving the distinctive beliefs of religion either from the data of physical science or by the methods of physical science.
>
> (The Nature of the Physical World, p. 333.)

The lack of finality of scientific theories would be a very serious limitation of our argument, if we had staked much on their permanence. The religious reader may well be content that I have not offered him a God revealed by the quantum theory, and therefore liable to be swept away in the next scientific revolution.

(Ibid. p. 353.)

It is probably true that the recent changes of scientific thought remove some of the obstacles to a reconciliation of religion with science, but this must be carefully distinguished from any proposal to base religion on scientific discovery. For my own part, I am wholly opposed to any such attempt.

(Science and the Unseen World, p. 45.)

The passages quoted by Mr. Cohen make it clear that I do not suggest that the new physics "proves religion" or indeed gives any positive grounds for religious faith. But it gives strong grounds for an idealistic philosophy which, I suggest, is hospitable towards a spiritual religion, it being understood that *the guest must provide his own credentials.* In short, the new conception of the physical universe puts me in a position to defend religion against a particular charge, viz. the charge of being incompatible with physical science. It is not a general panacea against atheism. If this is understood, . . . it explains my "great readiness to take the present standing of certain theories of physics as being final"; anybody can defend religion against science by speculating on the possibility that science may be mistaken. It explains why I sometimes take the essential truth of religion for granted; the soldier whose task is to defend one side of a fort must assume that the defenders of the other side have not been overwhelmed.

(*Article in* The Freethinker).

I now turn to the question, what must be put into the skeleton scheme of symbols. I have said that physical science stands aloof from this transmutation, and if I say anything positive on this side of the question it is *not as a scientist that I claim to speak.*

(*Broadcast Symposium,* Science and Religion).

The bearing of physical science on religion is that the scientist has, from time to time, assumed the duty of signalman and set up warnings of danger—not always unwisely. If I interpret the present situation rightly, a main-line signal which had been standing at danger has now been lowered. But nothing much is going to happen unless there is an engine.

[Eddington's point, as the following sections will make much clearer, is that physics—classical or quantum—can in no way offer *positive* support or even encouragement for a religious-mystical worldview. It is simply that, whereas classical physics was theoretically *hostile* to religion, modern physics is simply *indifferent* to it—it leaves so many theoretical holes in the universe that you may (or may not) fill them with religious substance, but if you do, it must be on philosophic or religious grounds. Physics cannot help you in the least, but it no longer objects to your efforts. This is what Eddington meant by, "If I interpret the present situation rightly, a main-line signal which had been standing at danger has now been lowered. But nothing much is going to happen unless there is an engine." Physics does not support mysticism, but it no longer denies it, and that, Eddington felt, opened a philosophic door to Spirit—but mysticism, not physics, must provide the "engine."

Eddington's view, which I fully endorse, would indeed be extremely good news—there is no longer any major physical-theoretical objection to spiritual realities—had not the new-age writers promised us the moon with "proofs" of mysticism *from* physics. Many people are therefore disappointed or let down by the apparently thin or weak nature of Eddington's pronouncement, whereas, in fact, this view—which is supported by virtually every theorist in this volume—is probably the strongest and most revolutionary conclusion *vis à vis* religion that has ever been "officially" advanced by theoretical science itself. It is a monumental and epochal turning point in science's stance towards religion; it seems highly unlikely it will ever be reversed, since it is logical and not empirical in nature (or *a priori* and not *a posteriori*); therefore, it, in all likelihood, marks final closure on that most nagging aspect of the age-old debate between the physical sciences and religion (or the geist-sciences). What more could one possibly want?]

Limitations of Physical Knowledge. Whenever we state the properties of a body in terms of physical qualities we are imparting knowledge as to the response of various metrical indicators to its presence, *and nothing more*. After all, knowledge of this kind is fairly comprehensive. A knowledge of the response of all kinds of objects—weighing-machines and other indicators—would determine completely its relation to its environment, leaving only its inner un-get-atable nature undetermined. In the relativity theory, we accept this as full knowledge, the nature of an object insofar as it is ascertainable by scientific inquiry being the abstraction of its relations to all surrounding objects. The progress of the relativity theory has been largely due to the development of a powerful mathematical calculus for dealing compendiously with an infinite scheme of pointer readings, and the technical term *tensor* used so largely in treatises on Einstein's theory may be translated *schedule of pointer readings*. It is part of the aesthetic appeal of the mathematical theory of relativity that the mathematics is so closely adapted to the physical conceptions. It is not so in all subjects. For example, we may admire the triumph of patience of the mathematician in predicting so closely the positions of the moon, but aesthetically the lunar theory is atrocious; it is obvious that the moon and the mathematician use different methods of finding the lunar orbit. But by the use of tensors the mathematical physicist precisely describes the nature of his subject-matter as a schedule of indicator readings, and those accretions of images and conceptions which have no place in physical science are automatically dismissed.

The recognition that our knowledge of the objects treated in physics consists solely of readings of pointers and other indicators transforms our view of the status of physical knowledge in a fundamental way. Until recently it was taken for granted that we had knowledge of a much more intimate kind of the entities of the external world. Let me give an illustration which takes us to the root of the great problem of the relations of matter and spirit. Take the living human brain endowed with mind and thought. Thought is one of the indisputable facts of the world. I know that I think, with a certainty which I cannot attribute to any of my physical knowledge of the world. More hypothetically, but on fairly plausible evidence, I am convinced that you have minds which think. Here then is a world fact to be investigated. The physicist brings his tools and commences systematic exploration. All that he discovers is a collection of atoms and electrons and fields of force arranged in space and time, apparently similar to those found in inorganic objects. He may trace other physical characteristics, energy, temperature, entropy. None of these is identical with thought. He might set down thought as an illusion—some perverse interpretation of the interplay of the physical entities that he has found. Or, if he sees the folly of calling the most undoubted element of our experience an illusion, he will have to face the tremendous question: How can this collection of ordinary atoms be a thinking machine? But what knowledge have we of the nature of atoms which renders it at all incongruous that they should constitute a thinking object? The Victorian physicist felt that he knew just what he was talking about when he used such terms as *matter* and *atoms*. Atoms were tiny billiard balls, a crisp statement that was supposed to tell you all about their nature in a way which could never be achieved for transcendental things like consciousness, beauty, or humour. But now we realise that science has nothing to say as to the intrinsic nature of the atom. The physical atom is, like everything else in physics, a schedule of pointer readings.

In science we study the linkage of pointer readings with pointer readings. The terms link together in endless cycle with the same inscrutable nature running through the whole. *There is nothing to prevent the assemblage of atoms constituting a brain from being of itself a thinking object in virtue of that nature which physics leaves undetermined and undeterminable.*

Cyclic Method of Physics. I must explain this reference to an endless cycle of physical terms. I will refer to Einstein's law of gravitation. This time I am going to expound it in a way so complete that there is not much likelihood that anyone will understand it. Never mind. We are not seeking further light on the cause of gravitation; we are interested in seeing what would really be involved in a *complete* explanation of anything physical.

Einstein's law, in its analytical form, is a statement that in empty space certain quantities called *potentials* obey certain lengthy differential equations. We make a memorandum of the word "potential" to remind us that we must later on explain what it means. We might conceive a world in which the potentials at every moment and every place had quite arbitrary values. The actual world is not so unlimited, the potentials being restricted to those values which conform to Einstein's equations. The next question is: What are potentials? They can be defined as quantities derived by quite simple mathematical calculations from certain fundamental quantities called *intervals*. (*mem.* Explain "interval.") If we know the values of the various intervals throughout the world, definite rules can be given for deriving the values of the potentials. What are intervals? They are relations between pairs of events which can be measured with a *scale* or a *clock* or with both. (*mem.* Explain "scale" and "clock.") Instructions can be given for the correct use of the scale and clock so that the interval is given by a prescribed combination of their readings. What are scales and clocks? A scale is a graduated strip of *matter* which . . . (*mem.* Explain "matter.") On second thoughts, I will leave the rest of the description as "an exercise to the reader" since it would take rather a long time to enumerate all the properties and niceties of behaviour of the material standard which a physicist would accept as a perfect scale or a perfect clock. We pass on to the next question: What is matter? We have dismissed the metaphysical conception of substance. We might perhaps here describe the atomic and electrical structure of matter, but that leads to the microscopic aspects of the world, whereas we are here taking the macroscopic outlook. Confining ourselves to mechanics, which is the subject in which the law of gravitation arises, matter may be defined as the embodiment of three related physical quantities, *mass* (or energy), *momentum,* and *stress.* What are "mass," "momentum," and "stress?" It is one of the most far-reaching achievements of Einstein's theory that it has given an exact answer to this question. They are rather formidable looking expressions containing the *potentials* and their first and second derivatives with respect to the coordinates. What are the potentials? Why, that is just what I have been explaining to you!

The definitions of physics proceed according to the method immortalised in "The House that Jack Built": This is the potential, that was derived from the interval, that was measured by the scale, that was made from the matter, that embodied the stress, that . . . But instead of finishing with Jack, whom, of course, every youngster must know without need for an introduction, we make a circuit back to the beginning of the rhyme: . . . that worried the cat, that killed the rat, that ate the malt, that lay in the house, that was built by the priest all shaven and shorn, that married the man. . . . Now we can go round and round forever.

But perhaps you have already cut short my explanation of gravitation. When we reached *matter* you had had enough of it. "Please do not explain any more. I happen to know what matter is." Very well; matter is something that Mr. X knows. Let us see how it goes: This is the potential that was derived from the interval that was measured by the scale that was made from the matter that Mr. X knows. Next question: What is Mr. X?

Well, it happens that physics is not at all anxious to pursue the question: What is Mr. X? It is not disposed to admit that its elaborate structure of a physical universe is "The House that Mr. X Built." It looks upon Mr. X—and, more particularly, the part of Mr. X that *knows*—as a rather troublesome tenant who, at a late stage of the world's history, has come to inhabit a structure which inorganic Nature has, by slow evolutionary progress, contrived to build. And so it turns aside from the avenue leading to Mr. X—and beyond—and closes up its cycle leaving him out in the cold.

From its own point of view, physics is entirely justified. That matter, in some indirect way, comes within the purview of Mr. X's mind is not a fact of any utility for a theoretical scheme of physics. We cannot embody it in a differential equation. It is ignored, and the physical properties of matter and other entities are expressed by their linkages in the cycle. And you can see how by the ingenious device of the cycle physics secures for itself a self-contained domain for study with no loose ends projecting into the unknown. All other physical definitions have the same kind of interlocking. Electric force is defined as something which causes motion of an electric charge; an electric charge is something that exerts something that produces motion of something that exerts something that produces . . . *ad infinitum*.

To know what there is about Mr. X which makes him behave in this strange way, we must look not to a physical system of inference, *but to that insight beneath the symbols which, in our own minds, we possess*. It is by this insight that we can finally reach an answer to our question: What is Mr. X?

So long as physics, in tinkering with the familiar world, was able to retain those aspects which appeal to the aesthetic side of our nature, it might with some show of reason make claim to cover the whole of experience; those who claimed that there was another, religious aspect of our

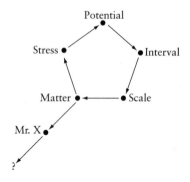

existence had to fight for their claim. But now that its picture omits so much that is obviously significant, there is no suggestion that it is the whole truth about experience. To make such a claim would bring protest not only from the religiously minded, but from all who recognise that Man is not merely a scientific measuring machine.

Physics provides a highly perfected answer to one specialised problem which confronts us in experience. I do not wish to minimise the importance of the problem and the value of the solution. In order to focus the problem, the various faculties of the observer have been discarded, and even his sensory equipment simplified, until the problem becomes such as our methods are adequate to solve. For the physicist, the observer has become a symbol dwelling in a world of symbols. But before ever we handed over the problem to the physicist, we had a glimpse of Man as a spirit in an environment akin to his own spirit.

Insofar as I refer in these lectures to an experience reaching beyond the symbolic equations of physics, I am not drawing on any specialised scientific knowledge; I depend, as anyone might do, on that which is the common inheritance of human thought.

We recognise that the type of knowledge after which physics is striving is much too narrow and specialised to constitute a complete understanding of the environment of the human spirit. A great many aspects of our ordinary life and activity take us outside the outlook of physics. For the most part, no controversy arises as to the admissibility and importance of these aspects; we take their validity for granted and adapt our life to them without any deep self-questioning. Any discussion as to whether they are compatible with the truth revealed by physics is purely academic; whatever the outcome of the discussion, we are not likely to sacrifice them, knowing as we do at the outset that the nature of Man would be incomplete without such outlets. It is, therefore, somewhat of an anomaly that among the many extraphysical aspects of experience, religion alone should be singled out as specially in need of reconciliation with the knowledge contained in science: Why should anyone suppose that all that matters to human nature can be assessed with a measuring rod or expressed in terms of the intersections of world-lines? If defence is needed, the defence of a religious outlook must, I think, take the same form as the defence of an aesthetic outlook. The sanction seems to lie in an inner feeling of growth or achievement found in the exercise of the aesthetic faculty and equally in the exercise of the religious faculty. It is akin to the inner feeling of the scientist which persuades him that, through the exercise of another faculty of the mind, namely its reasoning power, we reach something after which the human spirit is bound to strive.

It is by looking into our own nature that we first discover the failure of the physical universe to be co-extensive with our experience of reality. The "something to which truth matters" must surely have a place in reality whatever definition of reality we may adopt. In our own nature, or through the contact of our consciousness with a nature transcending ours, there are other things that claim the same kind of recognition—a sense of beauty, of morality, and finally, at the root of all spiritual religion, an experience which we describe as the presence of God. In suggesting that these things constitute a spiritual world, I am not trying to substantialise them or objectivise them—to make them out other than we find them to be in our experience of them. But I would say that when from the human heart, perplexed with the mystery of existence, the cry goes up, "What is it all about?" it is no true answer to look only at that part of experience which comes to us through certain sensory organs and reply: "It is about atoms and chaos; it is about a universe of fiery globes rolling on to impending doom; it is about tensors and non-commutative algebra." Rather, it is about a spirit in which truth has its shrine, with potentialities of self-fulfillment in its response to beauty and right. Shall I not also add that even as light and colour and sound come into our minds at the prompting of a world beyond, so these other stirrings of consciousness come from something which, whether we describe it as beyond or deep within ourselves, is greater than our own personality?

It is the essence of religion that it presents this side of experience as a matter of everyday life. To live in it, we have to grasp it in the form of familiar recognition and not as a series of abstract scientific statements. The man who commonly spoke of his ordinary surroundings in scientific language would be insufferable. If God means anything in our daily lives, I do not think we should feel any disloyalty to truth in speaking and thinking of him unscientifically, any more than in speaking and thinking unscientifically of our human companions.

The Definition of Reality. It is time we came to grips with the loose terms Reality and Existence, which we have been using without any inquiry into what they are meant to convey. I am afraid of this word Reality, not connoting an ordinarily definable characteristic of the things it is applied to but used as though it were some kind of celestial halo. It is, of course, possible to obtain consistent use of the word "reality" by adopting a conventional definition. My own practice would probably be covered by the definition that a thing may be said to be real if it is the goal of a type of inquiry to which I personally attach importance. But if I insist on no more than this I am whittling down the significance that is generally assumed. In physics, we can give a cold scientific definition of reality which is free from all sentimental mystification. But this is not quite fair play, because the word "reality" is generally used *with the intention of evoking sentiment.* It is a grand word for a peroration. "The right honourable speaker went on to declare that the concord and amity for which he had unceasingly striven had now become a reality (loud cheers)." The conception which it is so troublesome to apprehend is not "reality" but "reality (loud cheers)."

Let us first examine the definition according to the purely scientific usage of the word, although it will not take us far enough. The only subject presented to me for study is the content of my consciousness. You are able to communicate to me part of the content of your consciousness which thereby becomes accessible in my own. For reasons which are generally admitted, though I should not like to have to prove that they are conclusive, I grant your consciousness equal status with my own; and I use this secondhand part of my consciousness to "put myself in your place." Accordingly, my subject of study becomes differentiated into the contents of many consciousnesses, each content constituting a *viewpoint.* There then arises the problem of combining the viewpoints, and it is through this that the external world of physics arises. Much that is in any one consciousness is individual, much is apparently alterable by volition, but there is a stable element which is common to other consciousnesses. That common element we desire to study, to describe as fully and accurately as possible, and to discover the laws by which it combines now with one viewpoint, now with another. This common element cannot be placed in one man's consciousness rather than in another's; it must be in neutral ground—an external world.

If we are to find for the atoms and electrons of the external world not merely a conventional reality but "reality (loud cheers)" we must look not to the end but to the beginning of the quest. It is at the beginning that we must find that sanction which raises these entities above the mere products of an arbitrary mental exercise. This involves some kind of assessment of the impulse which sets us forth on the voyage of discovery. How can we make such assessment? Not by any reasoning that I know of. Reasoning would only tell us that the impulse might be judged by the success of the adventure—whether it leads in the end to things which really exist and wear the halo in their own right; it takes us to and fro like a shuttle along the chain of inference in vain search for the elusive halo. But, legitimately or not, the mind is confident that it can distinguish certain quests as sanctioned by indisputable authority. We may put it in different ways; the impulse to this quest is part of our very nature; it is the expression of a purpose which has possession of us. Is this precisely what we meant when we sought to affirm the reality of the external world? It goes some way toward giving it a meaning but is scarcely the full equivalent. I doubt if we really satisfy the conceptions behind that demand unless we make the bolder hypothesis that the quest and all that is reached by it are of worth in the eyes of an Absolute Valuer.

Whatever justification at the source we accept to vindicate the reality of the external world, it can scarcely fail to admit on the same footing much that is outside physical science. Although no long chains or regularised inference depend from them, we recognise that other fibres of our being extend in directions away from sense-impressions. I am not greatly concerned to borrow words like "existence" and "reality" to crown these other departments of the soul's interest. I would rather put it that any raising of the question of reality in its transcendental sense (whether the question emanates from the world of physics or not) leads us to a perspective from which we see man not as a bundle of sensory impressions, but conscious of purpose and responsibilities to which the external world is subordinate.

From this perspective we recognise a spiritual world alongside the physical world. Experience—that is to say, the self *cum* environment—comprises more than can be embraced in the physical world, restricted as it is to a complex of metrical symbols. The physical world is, we have seen, the answer to one definite and urgent problem arising in a survey of experience; no other problem has been followed up with anything like the same precision and elaboration. Progress toward an understanding of the non-sensory constituents of our nature is not likely to follow similar lines and, indeed, is not animated by the same aims. If it is felt that this difference is so wide that the phrase spiritual *world* is a misleading analogy, I will not insist on the term. All I would claim is that *those who in the search for truth start from consciousness as a seat of self-knowledge with interests and responsibilities not confined to the material plane* are just as much facing the hard facts of experience as those who start from consciousness as a device for reading the indications of spectroscopes and micrometers.

What is the ultimate truth about ourselves? Various answers suggest themselves. We are a bit of stellar matter gone wrong. We are physical machinary—puppets that strut and talk and laugh and die as the hand of time pulls the strings beneath. But there is one elementary inescapable answer. *We are that which asks the question.* Whatever else there may be in our nature, responsibility towards truth is one of its attributes. This side of our nature is aloof from the scrutiny of the physicist. I do not think it is sufficiently covered by admitting a mental aspect of our being. It has to do with conscience rather than with consciousness. Concern with truth is one of those things which make up the spiritual nature of Man. There are other constituents of our spiritual nature which are perhaps as self-evident, but it is not so easy to force an admission of their existence. We cannot recognise a problem of experience without at the same time recognising ourselves as truth-seekers involved in the problem. The strange association of soul and body—of responsibility toward truth with a particular group of carbon compounds—is a problem in which we naturally feel intense interest, but it is not an anxious interest, as though the existence of a spiritual significance of experience were hanging in the balance. That significance is to be regarded rather as a datum of the problem; the solution must fit the data; we must not alter the data to fit an alleged solution.

It would be foolish to deny the magnitude of the gulf between our understanding of the most complex form of inorganic matter and the simplest form of life. Let us suppose, however, that some day this gulf is bridged, and science is able to show how from the entities of physics creatures might be formed which are counterparts of ourselves even to the point of being endowed with life. The scientist will perhaps point out the nervous mechanism of the creature, its powers of motion, of growth, of reproduction, and end by saying "That's you." But it has yet to satisfy the inescapable test. Is it concerned with truth as I am? Then I will acknowledge that it is indeed myself. The scientist might point to motions in the brain and say that these really mean sensations, emotions, thoughts, and perhaps supply a code to translate the motions into the corresponding thoughts. Even if we could accept this inadequate substitute for consciousness as we intimately know it, we must still protest: "You have shown us a creature which thinks and believes; you have not shown us a creature to whom it *matters* that what it thinks and believes should be true." The inmost ego, possessing what I have called the inescapable attribute, can never be part of the physical world unless we alter the meaning of the word "physical" so as to the synonymous with "spiritual"—a change scarcely to the advantage of clear thinking. But having disowned our supposed double, we can say to the scientist: "If you will hand over this Robot who pretends to be me, and let it be filled with the attribute at present lacking and perhaps other spiritual attributes which I claim as equally self-evident, we may arrive at something that is indeed myself."

A few years ago the suggestion of taking the physically constructed man and adapting him to a spiritual nature by casually adding something, would have been a mere figure of speech—a verbal gliding over of insuperable difficulties. In much the same way, we talk loosely of constructing a Robot and then breathing life into it. A Robot is presumably not constructed to bear such last-minute changes of design; it is a delicate piece of mechanism made to work mechanically, and to adapt it to anything else would involve entire reconstruction. To put it crudely, if you want to fill a vessel with anything you must make it hollow, and the old-fashioned material body was not hollow enough to be a receptacle of mental or of spiritual attributes. The result was to place consciousness in the position of an intruder in the physical world. We had to choose between explaining it away as an illusion or perverse misrepresentation of what was really going on in the brain, and admitting an extraneous agent which had power to suspend the regular laws of Nature and asserted itself by brute interference with the atoms and molecules in contact with it.

Our present conception of the physical world is *hollow* enough to hold almost anything. I think the reader will agree. There may indeed be a hint of ribaldry in his hearty assent. What we are dragging to light as the basis of all phenomena is a scheme of symbols connected by mathematical equations. That is what physical reality boils down to when probed by the methods which a physicist can apply. A skeleton scheme of symbols proclaims its own hollowness. It can be—nay it cries out to be—filled with something that shall transform it from skeleton into substance, from plan into execution, from symbols into an interpretation of the symbols. And if ever the physicist solves the problem of the living body, he should no longer be tempted to point to his result and say "That's you." He should say rather "That is the aggregation of symbols which stands for you in my description and explanation of those of your properties which I can observe and measure. If you claim a deeper insight into your own nature by which you can interpret these symbols—a more intimate knowledge of the reality which I can only deal with by symbolism—you can rest assured that I have no rival interpretation to propose." The skeleton is the contribution of physics to the solution of the Problem of Experience; from the clothing of the skeleton it stands aloof.

Let us now consider our answer to the question whether the nature of reality is material or spiritual or a combination of both. I have often indicated my dislike of the word "reality" which so often darkens counsel, but I state the question as it is commonly worded and answer what I think is in the mind of the querist.

I will first ask another question. Is the ocean composed of water or of waves or of both? Some of my fellow passengers on the Atlantic were emphatically of the opinion that it is composed of waves, but I think the ordinary unprejudiced answer would be that it is composed of water. At least if we declare our belief that the nature of the ocean is aqueous, it is not likely that anyone will challenge us and assert that on the contrary its nature is undulatory, or that it is a dualism part aqueous and part undulatory. Similarly, I assert that the nature of all reality is spiritual, not material nor a dualism of matter and spirit. The hypothesis that its nature can be, to any degree, material does not enter into my reckoning, because as we now understand matter, the putting together of the adjective "material" and the noun "nature" does not make sense .

Interpreting the term material (or more strictly, physical) in the broadest sense as that with which we can become acquainted through sensory experience of the external world, we recognise now that it corresponds to the waves, not to the water of the ocean of reality. My answer does not deny the existence of the physical world, any more than the answer that the ocean is made of water denies the existence of ocean waves; only we do not get down to the intrinsic nature of things that way. Like the symbolic world of physics, a wave is a conception which is hollow enough to hold almost anything; we can have waves of water, of air, of aether, and (in quantum theory) waves of probability. So after physics has shown us the waves, we have still to determine the content of the waves by some other avenue of knowledge. If you will understand that the spiritual aspect of experience is to the physical aspect in the same kind of relation as the water to the wave form, I can leave you to draw up your own answer to the question propounded at the beginning of this section and so avoid any verbal misunderstanding. What is more important, you will see how easily the two aspects of experience now dovetail together, not contesting each other's place. It is almost as though the modern conception of the physical world had deliberately left room for the reality of spirit and consciousness.

The elements of consciousness are particular thoughts and feelings; the elements of the brain cell are atoms and electrons. But the two analyses do not run parallel to one another. Whilst, therefore, I contemplate a spiritual domain underlying the physical world as a whole, I do not think of it as distributed so that to each element of time and space there is a corresponding portion of the spiritual background. My conclusion is that, although for the most part our inquiry into the problem of experience ends in a veil of symbols, there is an immediate knowledge in the minds of conscious beings which lifts the veil in places; what we discern through these openings is of mental and spiritual nature. Elsewhere we see no more than the veil.

It is probably true that the recent changes of scientific thought remove some of the obstacles to a reconciliation of religion with science, but this must be carefully distinguished from any proposal to base religion on scientific discovery. For my own part, I am wholly opposed to any such attempt. Briefly, the position is this. We have learnt that the exploration of the external world by the methods of physical science leads not to a concrete reality but to a shadow world of symbols, beneath which those methods are unadapted for penetrating. If to-day you ask a physicist what he has finally made out the aether or the electron to be, the answer will not be a description in terms of billiard balls or fly-wheels or anything concrete; he will point instead to a number of symbols and a set of mathematical equations which they satisfy. What do the symbols stand for? The mysterious reply is given that physics is indifferent to that; it has no means of probing beneath the symbolism. To understand the phenomena of the physical world, it is necessary to know the equations which the symbols obey but not the nature of that which is being symbolised.

Feeling that there must be more behind, we return to our starting point in human consciousness—the one centre where more might become known. There we find other stirrings, other revelations (true or false) than those conditioned by the world of symbols.

We all share the strange delusion that a lump of matter is something whose general nature is easily comprehensible whereas the nature of the human spirit is unfathomable. But consider how our supposed acquaintance with the lump of matter is attained. Some influence emanating from it plays on the extremity of a nerve, starting a series of physical and chemical changes which are propagated along the nerve to a brain-cell; there a mystery happens, and an image or sensation arises in the mind which cannot purport to resemble the stimulus which excites it. Everything known about the material world must in one way or another have been inferred from these stimuli transmitted along the nerves. It is an astonishing feat of deciphering that we should have been able to infer an orderly scheme of natural knowledge from such indirect communication. But clearly there is one kind of knowledge which cannot pass through such channels, namely knowledge of the intrinsic nature of that which lies at the far end of the line of communication. The inferred knowledge is a skeleton frame, the entities which build the frame being of undisclosed nature. For that reason, they are described by symbols, as the symbol x in algebra stands for an unknown quantity.

The mind as a central receiving station reads the dots and dashes of the incoming nerve-signals. By frequent repetition of their call-signals the various transmitting stations of the outside world become familiar. We begin to feel quite a homely acquaintance with $_2$LO and $_5$XX. But a broadcasting station is not *like* its call-signal; there is no commensurability in their nature. So too the chairs and tables around us which broadcast to us incessantly those signals which affect our sight and touch cannot in their nature be like unto the signals or to the sensations which the signals awake at the end of their journey.

Penetrating as deeply as we can by the methods of physical investigation into the nature of a human being we reach only symbolic description. Far from attempting to dogmatise as to the nature of the reality thus symbolised, physics most strongly insists that its methods do not penetrate behind the symbolism. Surely then that mental and spiritual nature of ourselves, known in our minds by an intimate contact transcending the methods of physics, supplies just that interpretation of the symbols which science is admittedly unable to give. It is just because we have a real and not merely a symbolic knowledge of our own nature that our nature seems so mysterious; we reject as inadequate that merely symbolic description which is good enough for dealing with chairs and tables and physical agencies that affect us only by remote communication.

In comparing the certainty of things spiritual and things temporal, let us not forget this: mind is the first and most direct thing in our experience; all else is remote inference.

That environment of space and time and matter, of light and colour and concrete things, which seems so vividly real to us is probed deeply by every device of physical science and at the bottom we reach symbols. Its substance has melted into shadow. Nonetheless, it remains a real world if there is a background to the symbols—an unknown quantity which the mathematical symbol x stands for. We think we are not wholly cut off from this background. It is to this background that our own personality and consciousness belong, and those spiritual aspects of our nature not to be described by any symbolism or at least not by symbolism of the numerical kind to which mathematical physics has hitherto restricted itself. Our story of evolution ended with a stirring in the brain-organ of the latest of Nature's experiments, but that stirring of consciousness transmutes the whole story and gives meaning to its symbolism. Symbolically, it is the end, but, looking behind the symbolism, it is the beginning.

20

Mind-Stuff

I WILL TRY TO BE AS DEFINITE AS I CAN as to the glimpse of reality which we seem to have reached. Only I am well aware that in committing myself to details I shall probably blunder. Even if the right view has here been taken of the philosophical trend of modern science, it is premature to suggest a cut-and-dried scheme of the nature of things. If the criticism is made that certain aspects are touched on which come more within the province of the expert psychologist, I must admit its pertinence. The recent tendencies of science do, I believe, take us to an eminence from which we can look down into the deep waters of philosophy; if I rashly plunge into them, it is not because I have confidence in my powers of swimming, but to try to show that the water is really deep.

To put the conclusion crudely—the stuff of the world is mind-stuff. As is often the way with crude statements, I shall have to explain that by "mind" I do not here exactly mean mind and by "stuff" I do not at all mean stuff. Still, this is about as near as we can get to the idea in a simple phrase. The mind-stuff of the world is, of course, something more general than our individual conscious minds, but we may think of its nature as not altogether foreign to the feelings in our consciousness. The realistic matter and fields of force of former physical theory are altogether irrelevant—except in so far as the mind-stuff has itself spun these imaginings. The symbolic matter and fields of force of present-day theory are more relevant, but they bear to it the same relation that the bursar's accounts bear to the activity of the college. Having granted this, the mental activity of the part of the world constituting ourselves occasions no surprise; it is known to us by direct self-knowledge, and we do not explain it away as something other than we know it to be—or, rather, it knows itself to be. It is the physical aspects of the world that we have to explain. Our bodies are more mysterious than our minds—at least they would be, only that we can set the mystery on one side by the device of the cyclic scheme of physics, which enables us to study their phenomenal behaviour without ever coming to grips with the underlying mystery.

The mind-stuff is not spread in space and time; these are part of the cyclic scheme ultimately derived out of it. But we must presume that in some other way or aspect it can be differentiated into parts. Only here and there does it rise to the level of consciousness, but from such islands proceeds all knowledge. Besides the direct knowledge contained in each self-knowing unit, there is inferential knowledge. The latter includes our knowledge of the physical world. It is necessary to keep reminding ourselves that all knowledge of our environment from which the world of physics is constructed, has entered in the form of messages transmitted along the nerves to the seat of consciousness. Obviously, the messages travel in code. When messages relating to a table are travelling in the nerves, the nerve-disturbance does not in the least resemble either the external table that originates the mental impression or the conception of the table that arises in consciousness. In the central clearing station the incoming messages are sorted and decoded, partly by instinctive image-building inherited from the experience of our ancestors, partly by scientific comparison and reasoning. By this very indirect and hypothetical inference all our supposed acquaintance with and our theories of a world outside us have been built up. We are acquainted with an external world because its fibres run into our consciousness; it is only our own ends of the fibres that we actually know; from those ends, we more or less successfully reconstruct the rest, as a palaeontologist reconstructs an extinct monster from its footprint.

The mind-stuff is the aggregation of relations and relata which form the building material for the physical world. Our account of the building process shows, however, that much that is implied in the relations is dropped as unserviceable for the required building. Our view is practically that urged in 1875 by W. K. Clifford: "The succession of feelings which constitutes a man's consciousness is the reality which produces in our minds the perception of the motions of his brain."

That is to say, that which the man himself knows as a succession of feelings is the reality which when probed by the appliances of an outside investigator affects their readings in such a way that it is identified as a configuration of brain-matter. Again Bertrand Russell writes:

What the physiologist sees when he examines a brain is in the physiologist, not in the brain he is examining. What is in the brain by the time the physiologist examines it if it is dead, I do not profess to know; but while its owner was alive, part, at least, of the contents of his brain consisted of his percepts, thoughts, and feelings. Since his brain also consisted of electronics, we are compelled to conclude that an electron is a grouping of events, and that if the electron is in a human brain, some of the events composing it are likely to be some of the "mental states" of the man to whom the brain belongs. Or, at any rate, they are likely to be parts of such "mental states"—for it must not be assumed that part of a mental state must be a mental state. I do not wish to discuss what is meant by a "mental state"; the main point for us is that the term must include percepts. Thus a percept is an event or a group of events, each of which belongs to one or more of the groups constituting the electrons in the brain. This, I think, is the most concrete statement that can be made about electrons; everything else that can be said is more or less abstract and mathematical.

I quote this partly for the sake of the remark that it must not be assumed that part of a mental state must necessarily be a mental state. We can, no doubt, analyse the content of consciousness during a short interval of time into more or less elementary constituent feelings, but it is not suggested that this psychological analysis will reveal the elements out of whose measure-numbers the atoms or electrons are built. The brain-matter is a partial aspect of the whole mental state, but the analysis of the brain-matter by physical investigation does not run at all parallel with the analysis of the mental state by psychological investigation. I assume that Russell meant to warn us that, in speaking of part of a mental state, he was not limiting himself to parts that would be recognised as such psychologically, and he was admitting a more abstract kind of dissection.

This might give rise to some difficulty if we were postulating complete identity of mind-stuff with consciousness. But we know that in the mind there are memories not in consciousness at the moment, but capable of being summoned into consciousness. We are vaguely aware that things we cannot recall are lying somewhere about and may come into the mind at any moment. Consciousness is not sharply defined, but fades into subconsciousness; beyond that, we must postulate something indefinite but yet continuous with our mental nature. This I take to be the world-stuff. We liken it to our conscious feelings because, now that we are convinced of the formal and symbolic character of the entities of physics, there is nothing else to liken it to.

It is sometimes urged that the basal stuff of the world should be called "neutral stuff" rather than "mind-stuff," since it is to be such that both mind and matter originate from it. If this is intended to emphasize that only limited islands of it constitute actual minds, and that even in these islands that which is known mentally is not equivalent to a complete inventory of all that may be there, I agree. In fact, I should suppose that the self-knowledge of consciousness is mainly or wholly a knowledge which eludes the inventory method of description. The term "mind-stuff" might well be amended, but neutral stuff seems to be the wrong kind of amendment. It implies that we have two avenues of approach to an understanding of its nature. *We have only one approach, namely, through our direct knowledge of mind. The supposed approach through the physical world leads only into the cycle of physics, where we run round and round like a kitten chasing its tail and never reach the world-stuff at all.*

I assume that we have left the illusion of substance so far behind that the word "stuff" will not cause any misapprehension. I certainly do not intend to materialise or substantialise mind. Mind is—but you know what mind is like, so why should I say more about its nature? The word "stuff" has reference to the function it has to perform as a basis of world-building and does not imply any modified view of its nature.

It is difficult for the matter-of-fact physicist to accept the view that the substratum of everything is of mental character. But no one can deny that mind is the first and most direct thing in our experience, and all else is remote interference—inference either intuitive or deliberate. Probably it would never have occurred to us (as a serious hypothesis) that the world could be based on anything else, had we not been under the impression that there was a rival stuff with a more comfortable kind of "concrete" reality—something too inert and stupid to be capable of forging an illusion. The rival turns out to be a schedule of pointer readings, and, though a world of symbolic character can well be constructed from it, this is a mere shelving of the inquiry into the nature of the world of experience.

This view of the relation of the material to the spiritual world perhaps relieves to some extent a tension between science and religion. Physical science has seemed to occupy a domain of reality which is self-sufficient, pursuing its course independently of and indifferent to that which a voice within us asserts to be a higher reality. We are jealous of such independence. We are uneasy that there should be an apparently self-contained world in which God becomes an unnecessary hypothesis. We acknowledge that the ways of God are inscrutable, but is there not still in the religious mind something of that feeling of the prophets of old, who called on God to assert his kingship and, by sign or miracle, proclaim that the forces of Nature are subject to his command? And yet if the scientist were to repent and admit that it was necessary to include among the agents controlling the stars and the electrons an omnipresent spirit to whom we trace the sacred things of consciousness, would there not be even graver apprehension? We should suspect an intention to reduce God to a system of differential equations, like the other agents which at various times have been introduced to restore order in the physical scheme. That fiasco at any rate is avoided. For the sphere of the differential equations of physics is the metrical cyclic scheme extracted out of the broader reality. However much the ramifications of the cycles may be extended by further scientific discovery, they cannot from their very nature trench on the background in which they have their being—their actuality. It is in this background that our own mental consciousness lies; and here, if anywhere, we may find a Power greater than but akin to consciousness. It is not possible for the controlling laws of the spiritual substratum, which insofar as it is known to us in consciousness is essentially non-metrical, to be analogous to the differential and other mathematical equations of physics which are meaningless unless they are fed with metrical quantities. So that the crudest anthropomorphic image of a spiritual deity can scarcely be so wide of the truth as one conceived in terms of metrical equations.

One day I happened to be occupied with the subject of "Generation of Waves by Wind." I took down the standard treatise on hydro-dynamics, and under that heading I read—

> The equations (12) and (13) of the preceding Art. enable us to examine a related question of some interest, viz. the generation and maintenance of waves against viscosity, by suitable forces applied to the surface.
>
> *If the external forces* $\hat{p}_{yy}$, $\hat{p}_{xy}$ be given multiples of $e^{ikx + at}$, where k and a are prescribed, the equations in question determine A and C, and thence, by (9) the value of η. Thus we find

$$\frac{\acute{p}_{yy}}{g\rho\eta} = \frac{(\alpha^2 + 2vk^2\alpha + \sigma^2) A - i(\sigma^2 + 2vkma) C}{gk(A - iC)},$$

$$\frac{\acute{p}_{xy}}{g\rho\eta} = \frac{\alpha}{gk} \cdot \frac{2ivk^2A + (\alpha + 2vk^2) C}{(A - iC)},$$

where σ^2 has been written $g\kappa + T'\kappa^3$ as before . . .

And so on for two pages. At the end, it is made clear that a wind of less than half a mile an hour will leave the surface unruffled. At a mile an hour the surface is covered with minute corrugations due to capillary waves which decay immediately the disturbing cause ceases. At two miles an hour the gravity waves appear. As the author modestly concludes, "Our theoretical investigations give considerable insight into the incipient stages of wave-formation."

On another occasion the same subject of "Generation of Waves by Wind" was in my mind; but this time another book was more appropriate, and I read—

> There are waters blown by changing winds to laughter
> And lit by the rich skies, all day. And after,
>   Frost, with a gesture, stays the waves that dance
> And wandering loveliness. He leaves a white
>   Unbroken glory, a gathered radiance,
> A width, a shining peace, under the night.

The magic words bring back the scene. Again we feel Nature drawing close to us, uniting with us, till we are filled with the gladness of the waves dancing in the sunshine, with the awe of the moonlight on the frozen lake. These were not moments when we fell below ourselves. We do not look back on them and say, "It was disgraceful for a man with six sober senses and a scientific understanding to let himself be deluded in that way. I will take Lamb's *Hydrodynamics* with me next time." It is good that there should be such moments for us. Life would be stunted and narrow if we could feel no significance in the world around us beyond that which can be weighed and measured with the tools of the physicist or described by the metrical symbols of the mathematician.

Of course, it was an illusion. We can easily expose the rather clumsy trick that was played on us. Aethereal vibrations of various wavelengths, reflected at different angles from the disturbed interface between air and water, reached our eyes, and by photoelectric action caused appropriate stimuli to travel along the optic nerves to a brain-centre. Here the mind set to work to weave an impression out of the stimuli. The incoming material was somewhat meagre, but the mind is a great storehouse of associations that could be used to clothe the skeleton. Having woven an impression, the mind surveyed all that it had made and decided that it was very good. The critical faculty was lulled. We ceased to analyse and were conscious only of the impression as a whole. The warmth of the air, the scent of the grass, the gentle stir of the breeze, combined with the visual scene in one transcendent impression, around us and within us. Associations emerging from their storehouse grew bolder. Perhaps we recalled the phrase "rippling laughter." Waves—ripples—laughter—gladness—the ideas jostled one another. Quite illogically, we were glad, though what there can possibly be to be glad about in a set of aethereal vibrations no sensible person can explain. A mood of quiet joy suffused the whole impression. The gladness in ourselves was in Nature, in the waves, everywhere. That's how it was.

It was an illusion. Then why toy with it longer? These airy fancies which the mind, when we do not keep it severely in order, projects into the external world should be of no concern to the earnest seeker after truth. Get back to the solid substance of things, to the material of the water moving under the pressure of the wind and the force of gravitation in obedience to the laws of hydrodynamics. But the solid substance of things is another illusion. It too is a fancy projected by the mind into the external world. We have chased the solid substance from the continuous liquid to the atom, from the atom to the electron, and there we have lost it. But at least, it will be said, we have reached something real at the end of the chase—the protons and electrons. Or, if the new quantum theory condemns these images as too concrete and leaves us with no coherent images at all, at least we have symbolic coordinates and momenta and Hamiltonian functions devoting themselves with single-minded purpose to ensuring that $qp - pq$ shall be equal to $ih/2\pi$.

I have tried to show that by following this course we reach a cyclic scheme which, from its very nature, can only be a partial expression of our environment. It is not reality but the skeleton of reality. "Actuality" has been lost in the exigencies of the chase. Having first rejected the mind as a worker of illusion we have in the end to return to the mind and say, "Here are worlds well and truly built on a basis more secure than your fanciful illusions. But there is nothing to make any one of them an actual world. Please choose one and weave your fanciful images into it. That alone can make it actual." We have torn away the mental fancies to get at the reality beneath, only to find that the reality of that which is beneath is bound up with its potentiality of awakening these fancies. It is because the mind, the weaver of illusion, is also the only guarantor of reality that reality is always to be sought at the base of illusion. Illusion is to reality as the smoke to the fire. I will not urge that hoary untruth "There is no smoke without fire". But it is reasonable to inquire whether, in the mystical illusions of man, there is not a reflection of an underlying reality.

To put a plain question: Why should it be good for us to experience a state of self-deception such as I have described? I think everyone admits that it is good to have a spirit sensitive to the influences of Nature, good to exercise an appreciative imagination and not always to be remorselessly dissecting our environment after the manner of the mathematical physicists. And it is good not merely in a utilitarian sense, but in some purposive sense necessary to the fulfillment of the life that is given us. It is not a dope which it is expedient to take from time to time so that we may return with greater vigour to the more legitimate employment of the mind in scientific investigation. Just possibly it might be defended on the ground that it affords to the non-mathematical mind in some feeble measure that delight in the external world which would be more fully provided by an intimacy with its differential equations. (Lest it should be thought that I have intended to pillory hydrodynamics, I hasten to say in this connection that I would not rank the intellectual (scientific) appreciation on a lower plane than the mystical appreciation; I know of passages written in mathematical symbols which in their sublimity might vie with Rupert Brooke's sonnet.) But I think you will agree with me that it is impossible to allow that the one kind of appreciation can adequately fill the place of the other. Then how can it be deemed good if there is *nothing* in it but self-deception? That would be an upheaval of all our ideas of ethics. It seems to me that the only alternatives are either to count all such surrender to the mystical contact of Nature as mischievous and ethically wrong, or to admit that in these moods we catch something of the true relation of the world to ourselves—a relation not hinted at in a purely scientific analysis of its content. I think the most ardent materialist does not advocate, or, at any rate, does not practice, the first alternative, therefore, I assume the second alternative, that there is some kind of truth at the base of the illusion.

But we must pause to consider the extent of the illusion. Is it a question of a small nugget of reality buried under a mountain of illusion? If that were so, it would be our duty to rid our minds of some of the illusion at least, and try to know the truth in purer form. But I cannot think there is much amiss with our appreciation of the natural scene that so impresses us. I do not think a being more highly endowed than ourselves would prune away much of what we feel. It is not so much that the feeling itself is at fault as that our introspective examination of it wraps it in fanciful imagery. If I were to try to put into words the essential truth revealed in the mystic experience, it would be that our minds are not apart from the world, and the feelings that we have of gladness and melancholy and our yet deeper feelings are not of ourselves alone, but are glimpses of a reality transcending the narrow limits of our particular consciousness—that the harmony and beauty of the face of Nature is, at root, one with the gladness that transfigures the face of man. We try to express much the same truth when we say that the physical entities are only an extract of pointer readings and beneath them is a nature continuous with our own. But I do not willingly put it into words or subject it to introspection. We have seen how in the physical world the meaning is greatly changed when we contemplate it as surveyed from without instead of, as it essentially must be, from within. By introspection we drag out the truth for external survey, but in the mystical feeling the truth is apprehended from within and is, as it should be, a part of ourselves.

SYMBOLIC KNOWLEDGE AND INTIMATE KNOWLEDGE

May I elaborate this objection to introspection? We have two kinds of knowledge which I call symbolic knowledge and intimate knowledge. I do not know whether it would be correct to say that reasoning is only applicable to symbolic knowledge, but the more customary forms of reasoning have been developed for symbolic knowledge only. The intimate knowledge will not submit to codification and analysis, or, rather, when we attempt to analyse it the intimacy is lost and it is replaced by symbolism.

For an illustration let us consider Humour. I suppose that humour can be analysed to some extent and the essential ingredients of the different kinds of wit classified. Suppose that we are offered an alleged joke. We subject it to scientific analysis as we would a chemical salt of doubtful nature, and perhaps after careful consideration of all its aspects we are able to confirm that it really and truly is a joke. Logically, I suppose, our next procedure would be to laugh. But it may certainly be predicted that as the result of this scrutiny we shall have lost all inclination we may ever have had to laugh at it. It simply does not do to expose the inner workings of a joke. The classification concerns a symbolic knowledge of humour which preserves all the characteristics of a joke except its laughableness. The real appreciation must come spontaneously, not introspectively. I think this is a not unfair analogy for our mystical feeling for Nature, and I would venture even to apply it to our mystical experience of God. There are some to whom the sense of a divine presence irradiating the soul is one of the most obvious things of experience. In their view, a man without this sense is to be regarded as we regard a man without a sense of humour. The absence is a kind of mental deficiency. We may try to analyse the experience as we analyse humour, and construct a theology, or it may be an atheistic philosophy, which shall put into scientific form what is to be inferred about it. But let us not forget that the theology is symbolic knowledge, whereas the experience is intimate knowledge. And as laughter cannot be compelled by the scientific exposition of the structure of a joke, so a philosophic discussion of the attributes of God (or an impersonal substitute) is likely to miss the intimate response of the spirit which is the central point of the religious experience.

21

Defense of Mysticism

A DEFENCE OF THE MYSTIC might run something like this. We have acknowledged that the entities of physics can from their very nature form only a partial aspect of the reality. How are we to deal with the other part? It cannot be said that that other part concerns us less than the physical entities. Feelings, purpose, values, make up our consciousness as much as sense impressions. We follow up the sense impressions and find that they lead into an external world discussed by science; we follow up the other elements of our being and find that they lead not into a world of space and time, but surely somewhere. If you take the view that the whole of consciousness is reflected in the dance of electrons in the brain, so that each emotion is a separate figure of the dance, then all features of consciousness alike lead into the external world of physics. But I assume that you have followed me in rejecting this view, and that you agree that consciousness as a whole is greater than those quasi-metrical aspects of it which are abstracted to compose the physical brain. We have then to deal with those parts of our being unamenable to metrical specification, that do not make contact—jut out, as it were—into space and time. By dealing with them, I do not mean make scientific inquiry into them. The first step is to give acknowledged status to the crude conceptions in which the mind invests them, similar to the status of those crude conceptions which constitute the familiar material world.

Our conception of the familiar table was an illusion. But if some prophetic voice had warned us that it was an illusion and therefore we had not troubled to investigate further we should never have found the scientific table. To reach the reality of the table we need to be endowed with sense organs to weave images and illusions about it. And so it seems

209

to me that the first step in a broader revelation to man must be the awakening of image building in connection with the higher faculties of his nature, so that these are no longer blind alleys but open out into a spiritual world—a world partly of illusion, no doubt, but in which he lives no less than in the world, also of illusion, revealed by the senses.

The mystic, if haled before a tribunal of scientists, might perhaps end his defence on this note. He would say: "The familiar material world of everyday conception, though lacking somewhat in scientific truth, is good enough to live in; in fact, the scientific world of pointer readings would be an impossible sort of place to inhabit. It is a symbolic world and the only thing that could live comfortably in it would be a *symbol*. But I am not a symbol; I am compounded of that mental activity which is, from your point of view, a nest of illusion, so that to accord with my own nature I have to transform even the world explored by my senses. But I am not merely made up of senses; the rest of my nature has to live and grow. I have to render account of that environment into which it has its outlet. My conception of my spiritual environment is not to be compared with your scientific world of pointer readings; it is an every-day world to be compared with the material world of familiar experience. I claim it as no more real and no less real than that. Primarily, it is not a world to be analysed, but a world to be lived in."

Granted that this takes us outside the sphere of exact knowledge, and that it is difficult to imagine that anything corresponding to exact science will ever be applicable to this part of our environment, the mystic is unrepentant. Because we are unable to render exact account of our environment, it does not follow that it would be better to pretend that we live in a vacuum.

If the defence may be considered to have held good against the first onslaught, perhaps the next stage of the attack will be an easy tolerance. "Very well. Have it your own way. It is a harmless sort of belief—not like a more dogmatic theology. You want a sort of spiritual playground for those queer tendencies in man's nature, which sometimes take possession of him. Run away and play then, but do not bother the serious people who are making the world go round." The challenge now comes not from the scientific materialism which professes to seek a natural explanation of spiritual power, but from the deadlier moral materialism which despises it. Few deliberately hold the philosophy that the forces of progress are related only to the material side of our environment, but few can claim that they are not more or less under its sway. We must not interrupt the "practical men," these busy moulders of history carrying us at ever-increasing pace towards our destiny as an ant-heap of humanity infesting the earth. But is it true in history that material forces have been the most potent factors? Call it of God, of the Devil, fanaticism, unreason, but do not underrate the power of the mystic. Mysticism may be fought as error or believed as inspired, but it is no matter for easy tolerance—

> We are the music-makers
>   And we are the dreamers of dreams
> Wandering by lone sea-breakers
>   And sitting by desolate streams;
> World-losers and world-forsakers,
>   On whom the pale moon gleams:
> Yet we are the movers and shakers
>   Of the world for ever, it seems.

REALITY AND MYSTICISM

But a defence before the scientists may not be a defence to our own self-questionings. We are haunted by the word *reality*. I have already tried to deal with the questions which arise as to the meaning of reality, but it presses on us so persistently that, at the risk of repetition, I must consider it once more from the standpoint of religion. A compromise of illusion and reality may be all very well in our attitude towards physical surroundings, but to admit such a compromise into religion would seem to be a trifling with sacred things. Reality seems to concern religious beliefs much more than any others. No one bothers as to whether there is a reality behind humour. The artist who tries to bring out the soul in his picture does not really care whether and in what sense the soul can be said to exist. Even the physicist is unconcerned as to whether atoms or electrons really exist; he usually asserts that they do, but, as we have seen, existence is there used in a domestic sense and no inquiry is made as to whether it is more than a conventional term. In most subjects (perhaps not excluding philosophy), it seems sufficient to agree on the things that we shall call real, and afterward try to discover what we mean by the word. And so it comes about that religion seems to be the one field of inquiry in which the question of reality and existence is treated as of serious and vital importance.

But it is difficult to see how such an inquiry can be profitable. When

Dr. Johnson felt himself getting tied up in argument over "Bishop Berkeley's ingenious sophistry to prove the non-existence of matter, and that everything in the universe is merely ideal," he answered, "striking his foot with mighty force against a large stone, till he rebounded from it, 'I refute it *thus.*' " Just what that action assured him of is not very obvious, but apparently he found it comforting. And today the matter-of-fact scientist feels the same impulse to recoil from these flights of thought back to something kickable, although he ought to be aware by this time that what Rutherford has left us of the large stone is scarcely worth kicking.

There is still the tendency to use "reality" as a word of magic comfort like the blessed word "Mesopotamia." If I were to assert the reality of the soul or of God, I should certainly not intend a comparison with Johnson's large stone—a patent illusion—or even with the p's and q's of the quantum theory—an abstract symbolism. Therefore, I have no right to use the word in religion for the purpose of borrowing on its behalf that comfortable feeling which (probably wrongly) has become associated with stones and quantum coordinates.

Scientific instincts warn me that any attempt to answer the question "What is real?" in a broader sense than that adopted for domestic purposes in science, is likely to lead to a floundering among vain words and high-sounding epithets. We all know that there are regions of the human spirit untrammelled by the world of physics. In the mystic sense of the creation around us, in the expression of art, in a yearning towards God, the soul grows upward and finds the fulfillment of something implanted in its nature. The sanction for this development is within us, a striving born with our consciousness or an Inner Light proceeding from a greater power than ours. Science can scarcely question this sanction, for the pursuit of science springs from a striving which the mind is impelled to follow, a questioning that will not be suppressed. Whether in the intellectual pursuits of science or in the mystical pursuits of the spirit, the light beckons ahead and the purpose surging in our nature responds. Can we not leave it at that? Is it really necessary to drag in the comfortable word "reality" to be administered like a pat on the back?

The starting point of belief in mystical religion is a conviction of significance or, as I have called it earlier, the sanction of a striving in the consciousness. This must be emphasised because appeal to intuitive conviction of this kind has been the foundation of religion through all ages and I do not wish to give the impression that we have now found something new and more scientific to substitute. I repudiate the idea of proving the distinctive beliefs of religion either from the data of physical science or by the methods of physical science. Presupposing a mystical religion based not on science but (rightly or wrongly) on a self-known experience accepted as fundamental, we can proceed to discuss the various criticisms which science might bring against it or the possible conflict with scientific views of the nature of experience equally originating from self-known data.

It is necessary to examine further the nature of the conviction from which religion arises; otherwise, we may seem to be countenancing a blind rejection of reason as a guide to truth. There is a hiatus in reasoning, we must admit, but it is scarcely to be described as a rejection of reasoning. There is just the same hiatus in reasoning about the physical world if we go back far enough. We can only reason from data and the ultimate data must be given to us by a non-reasoning process—a self-knowledge of that which is in our consciousness. To make a start we must be aware of something. But that is not sufficient; we must be convinced of the significance of the significance of that awareness. We are bound to claim for human nature that, either of itself or as inspired by a power beyond, it is capable of making legitimate judgments of significance. Otherwise, we cannot even reach a physical world.

Accordingly, the conviction which we postulate is that certain states of awareness in consciousness have at least equal significance with those which are called sensations. It is perhaps not irrelevant to note that time by its dual entry into our minds to some extent bridges the gap between sense impressions and these other states of awareness. Amid the latter must be found the basis of experience from which a spiritual religion arises. The conviction is scarcely a matter to be argued about, it is dependent on the forcefulness of the feeling of awareness.

But, it may be said, although we may have such a department of consciousness, may we not have misunderstood altogether the nature of that which we believe we are experiencing? That seems to me to be rather beside the point. In regard to our experience of the physical world we have very much misunderstood the meaning of our sensations. It has been the task of science to discover that things are very different from what they seem. But we do not pluck out our eyes because they persist in deluding us with fanciful colourings instead of giving us the plain truth about wavelength. It is in the midst of such misrepresentations of environment (if you must call them so) that we have to live. It is, however, a very one-sided view of truth which can find in the glorious colouring of our surroundings nothing but misrepresentation—which takes the environment to be all-important and the conscious spirit to be inessential. It is the aim of physical science, so far as its scope extends, to lay bare the fundamental structure underlying the world, but science has also to explain if it can, or else humbly to accept, the fact that from this world have arisen minds capable of transmuting the bare structure into the richness of our experience. It is not misrepresentation but rather achievement—the result perhaps of long ages of biological evolution—that we should have fashioned a familiar world out of the crude basis. It is a fulfillment of the purpose of man's nature. If likewise the spiritual world has been transmuted by a religious colour beyond anything implied in its bare external qualities, it may be allowable to assert with equal conviction that this is not misrepresentation but the achievement of a divine element in man's nature.

May I revert again to the analogy of theology with the supposed science of humour which (after consultation with a classical authority) I venture to christen "geloeology." Analogy is not convincing argument, but it must serve here. Consider the proverbial Scotchman with strong leanings towards philosophy and incapable of seeing a joke. There is no reason why he should not take high honours in geloeology and, for example, write an acute analysis of the differences between British and American humour. His comparison of our respective jokes would be particularly unbiased and judicial, seeing that he is quite incapable of seeing the point of either. But it would be useless to consider his views as to which was following the right development; for that he would need a sympathetic understanding—he would (in the phrase appropriate to the other side of my analogy) need to be *converted*. The kind of help and criticism given by the geloeologist and the philosophical theologian is to secure that there is method in our madness. The former may show that our hilarious reception of a speech is the result of a satisfactory dinner and a good cigar rather than a subtle perception of wit; the latter may show that the ecstatic mysticism of the anchorite is the vagary of a fevered body and not a transcendent revelation. But I do not think we should appeal to either of them to discuss the reality of the sense with which we claim to be endowed, nor the direction of its right development. That is a matter for our inner sense of values which we all believe in to some extent, though it may be a matter of dispute just how far it goes. If we have no such sense then it would seem that not only religion, but the physical world and all faith in reasoning totter in insecurity.

I have sometimes been asked whether science cannot now furnish an argument which ought to convince any reasonable atheist. I could no more ram religious conviction into an atheist than I could ram a joke into the Scotchman. The only hope of "converting" the latter is that through contact with merry-minded companions he may begin to realise that he is missing something in life which is worth attaining. Probably in the recesses of his solemn mind there exists inhibited the seed of humour, awaiting an awakening by such an impulse. The same advice would seem to apply to the propagation of religion; it has, I believe, the merit of being entirely orthodox advice.

We cannot pretend to offer proofs. *Proof* is an idol before whom the pure mathematician tortures himself. In physics, we are generally content to sacrifice before the lesser shrine of *Plausibility*. And even the pure mathematician—that stern logician—reluctantly allows himself some prejudgments; he is never quite convinced that the scheme of mathematics is flawless, and mathematical logic has undergone revolutions as profound as the revolutions of physical theory. We are all alike stumblingly pursuing an ideal beyond our reach. In science, we sometimes have convictions as to the right solution of a problem which we cherish but cannot justify; we are influenced by some innate sense of the fitness of things. So too there may come to us convictions of the spiritual sphere which our nature bids us hold to. I have given an example of one such conviction which is rarely if ever disputed—that surrender to the mystic influence of a scene of natural beauty is right and proper for a human spirit, although it would have been deemed an unpardonable eccentricity in the "observer" contemplated in earlier chapters. Religious conviction is often described in somewhat analogous terms as a surrender; it is not to be enforced by argument on those who do not feel its claim in their own nature.

I think it is inevitable that these convictions should emphasis a personal aspect of what we are trying to grasp. We have to build the spiritual world out of symbols taken from our own personality, as we build the scientific world out of the metrical symbols of the mathematician. If not, it can only be left ungraspable—an environment dimly felt in moments of exaltation, but lost to us in the sordid routine of life. To turn it into more continuous channels we must be able to approach the World-Spirit in the midst of our cares and duties in that simpler relation of spirit to spirit in which all true religion finds expression.

A tide of indignation has been surging in the breast of the matter-of-fact scientist and is about to be unloosed upon us. Let us broadly survey the defence we can set up.

I suppose the most sweeping charge will be that I have been talking what at the back of my mind I must know is only a well-meaning kind of nonsense. I can assure you that there is a scientific part of me that has often brought that criticism during some of the later chapters. I will not say that I have been half-convinced, but at least I have felt a homesickness for the paths of physical science where there are more or less discernible handrails to keep us from the worst morasses of foolishness. But however much I may have felt inclined to tear up this part of the discussion and confine myself to my proper profession of juggling with pointer readings, I find myself holding to the main principles. Starting from aether, electrons, and other physical machinery, we cannot reach conscious man and render count of what is apprehended in his consciousness. Conceivably, we might reach a human machine interacting by reflexes with its environment, but we cannot reach rational man morally responsible to pursue the truth as to aether and electrons or to religion. Perhaps it may seem unnecessarily portentous to invoke the latest developments of the relativity and quantum theories merely to tell you this, but that is scarcely the point. We have followed these theories because they contain the conceptions of modern science; it is not a question of asserting a faith that science must ultimately be reconcilable with an idealistic view, but of examining how, at the moment, it actually stands in regard to it. There was a time when the whole combination of self and environment which makes up experience seemed likely to pass under the dominion of a physics much more iron bound than it is now. That overweening phase, when it was almost necessary to ask the permission of physics to call one's soul one's own, is past. The change gives rise to thoughts which ought to be developed. Even if we cannot attain to much clarity of constructive thought, we can discern that certain assumptions, expectations, or fears are no longer applicable.

Is it merely a well-meaning kind of nonsense for a physicist to affirm this necessity for an outlook beyond physics? It is worse nonsense to deny it. Or, as that ardent relativist the Red Queen puts it, "You call that nonsense, but I've heard nonsense compared with which that would be as sensible as a dictionary."

For if those who hold that there must be a physical basis for everything hold that these mystical views are nonsense, we may ask: What, then, is the physical basis of nonsense? The "problem of nonsense" touches the scientist more nearly than any other moral problem. He may regard the distinction of sense and nonsense, of valid and invalid reasoning, must be accepted at the beginning of every scientific inquiry. Therefore, it may well be chosen for examination as a test case.

If the brain contains a physical basis for the nonsense which it thinks, this must be some kind of configuration of the entities of physics—not precisely a chemical secretion, but not essentially different from that kind of product. It is as though when my brain says 7 times 8 are 56 its machinery is manufacturing sugar, but when it says 7 times 8 are 65 the machinery has gone wrong and produced chalk. But who says the machinery has gone wrong? As a physical machine, the brain has acted according to the unbreakable laws of physics; so why stigmatise its action? This discrimination of chemical products as good or evil has no parallel in chemistry. We cannot assimilate laws of thought to natural laws; they are laws which *ought* to be obeyed, not laws which *must* be obeyed; the physicist must accept laws of thought before he accepts natural law. "Ought" takes us outside chemistry and physics. It concerns something which wants or esteems sugar, not chalk, sense, not nonsense. A physical machine cannot esteem or want anything; whatever is fed into it it will chew up according to the laws of its physical machinery. That which in the physical world shadows the nonsense in the mind affords no ground for its condemnation. In a world of aether and electrons, we might perhaps encounter *nonsense;* we could not encounter *damned nonsense.*

And so my own concern lest I should have been talking nonsense ends in persuading me that I have to reckon with something that could not possibly be found in the physical world.

Another charge launched against these lectures may be that of admitting some degree of supernaturalism, which in the eyes of many is the same thing as superstition. Insofar as supernaturalism is associated with the denial of strict causality, I can only answer that that is what the modern scientific development of the quantum theory brings us to. But probably the more provocative part of our scheme is the role allowed to mind and consciousness. Yet I suppose that our adversary admits consciousness as a fact and he is aware that, but for knowledge by consciousness, scientific investigation could not begin. Does he regard consciousness as supernatural? Then it is he who is admitting the supernatural. Or does he regard it as part of Nature? So do we. We treat it in what seems to be its obvious position as the avenue of approach to the reality and significance of the world, as it is the avenue of approach to all scientific knowledge of the world. Or does he regard consciousness as something which, unfortunately, has to be admitted, but which it is scarcely polite to mention? Even so, we humour him. We have associated consciousness with a background untouched in the physical survey of the world and have given the physicist a domain where he can go round in cycles without ever encountering anything to bring a blush to his cheek. Here, a realm of natural law is secured to him covering all that he has ever effectively occupied. And, indeed, it has been quite as much the purpose of our discussion to secure such a realm where scientific method may work unhindered, as to deal with the nature of that part of our experience which lies beyond it. This defence of scientific method may not be superfluous. The accusation is often made that, by its neglect of aspects of human experience evident to a wider culture, physical science has been overtaken by a kind of madness leading it sadly astray. It is part of our contention that there exists a wide field of research for which the methods of physics suffice, into which the introducton of these other aspects would be entirely mischievous.

A besetting temptation of the scientific apologist for religion is to take some of its current expressions and, after clearing away crudities of thought (which must necessarily be associated with anything adapted to the everyday needs of humanity), to water down the meaning until little is left that could possibly be in opposition to science or to anything else. If the revised interpretation had first been presented no one would have raised vigorous criticism; on the other hand, no one would have been stirred to any great spiritual enthusiasm. It is the less easy to steer clear of this temptation because it is necessarily a question of degree. Clearly, if we are to extract from the tenets of a hundred different sects any coherent view to be defended some at least of them must be submitted to a watering-down process. I do not know if the reader will acquit me of having succumbed to this temptation in the passages where I have touched upon religion, but I have tried to make a fight against it. Any apparent failure has probably arisen in the following way. We have been concerned with the borderland of the material and spiritual worlds as approached from the side of the former. From this side, all that we could assert of the spiritual world would be insufficient to justify even the palest brand of theology that is not too emaciated to have any practical influence on man's outlook. But the spiritual world as understood in any serious religion is, by no means, a colourless domain. Thus by calling this hinterland of science a spiritual world, I may seem to have begged a vital question, whereas I intended only a provisional identification. To make it more than provisional an approach must be made from the other side. I am unwilling to play the amateur theologian and examine this approach in detail. I have, however, pointed out that the attribution of religious colour to the domain must rest on inner conviction; I think we should not deny validity to certain inner convictions, which seem parallel with the unreasoning trust in reason which is at the basis of mathematics, with an innate sense of the fitness of things which is at the basis of the science of the physical world, and with an irresistible sense of incongruity which is at the basis of the justification of humour. Or perhaps it is not so much a question of asserting the validity of these convictions as of recognising their function as an essential part of our nature. We do not defend the validity of seeing beauty in a natural landscape; we accept with gratitude the fact that we are so endowed as to see it that way.

It will perhaps be said that the conclusion to be drawn from these arguments from modern science is that religion first became possible for a reasonable scientific man about the year 1927. If we must consider that tiresome person, the consistently reasonable man, we may point out that not merely religion but most of the ordinary aspects of life first became possible for him in that year. Certain common activities (e.g. falling in love) are, I fancy, still forbidden him. If our expectation should prove well founded that 1927 has seen the final overthrow of strict causality by Heisenberg, Rohr, Born, and others, the year will certainly rank as one of the greatest epochs in the development of scientific philosophy. But seeing that before this enlightened era men managed to persuade themselves that they had to mould their own material future notwithstanding the yoke of strict causality, they might well use the same *modus vivendi* in religion.

The conflict [between science and religion] will not be averted unless both sides confine themselves to their proper domain, and a discussion which enables us to reach a better understanding as to the boundary should be a contribution towards a state of peace. There is still plenty of opportunity for frontier difficulties; a particular illustration will show this.

A belief not, by any means, confined to the more dogmatic adherents of religion is that there is a future non-material existence in store for us. Heaven is nowhere in space, but it is in time. (All the meaning of the belief is bound up with the word *future;* there is no comfort in an assurance of bliss in some *former* state of existence.) On the other hand, the scientist declares that time and space are a single continuum, and the modern idea of a Heaven in time but not in space is, in this respect, more at variance with science than the pre-Copernican idea of a Heaven above our heads. The question I am now putting is not whether the theologian or the scientist is right, but which is trespassing on the domain of the other? Cannot theology dispose of the destinies of the human soul in a non-material way without trespassing on the realm of science? Cannot science assert its conclusions as to the geometry of the space-time continuum without trespassing on the realm of theology? According to the assertion above, science and theology can make what mistakes they please provided that they make them *in their own territory;* they cannot quarrel if they keep to their own realms. But it will require a skillful drawing of the boundary line to frustrate the development of a conflict here.

The philosophic trend of modern scientific thought differs markedly from the views of thirty years ago. Can we guarantee that the next thirty years will not see another revolution, perhaps even a complete reaction? We may certainly expect great changes, and by that time many things will appear in a new aspect. That is one of the difficulties in the relations of science and philosophy; that is why the scientist, as a rule, pays so little heed to the philosophical implications of his own discoveries. By dogged endeavour, he is slowly and tortuously advancing to purer and purer truth, but his ideas seem to zigzag in a manner most disconcerting to the onlooker. Scientific discovery is like the fitting together of the pieces of a great jigsaw puzzle; a revolution of science does not mean that the pieces already arranged and interlocked have to be dispersed; it means that in fitting on fresh pieces we have had to revise our impression of what the puzzle-picture is going to be like. One day you ask the scientist how he is getting on; he replies, "Finely. I have very nearly finished this piece of blue sky." Another day you ask how the sky is progressing and are told, "I have added a lot more, but it was sea, not sky; there's a boat floating on the top of it." Perhaps next time it will have turned out to be a parasol upside down, but our friend is still enthusiastically delighted with the progress he is making. The scientist has his guesses as to how the finished picture will work out; he depends largely on these in his search for other pieces to fit, but his guesses are modified from time to time by unexpected developments as the fitting proceeds. These revolutions of thought as to the final picture do not cause the scientist to lose faith in his handiwork, for he is aware that the completed portion is growing steadily. Those who look over his shoulder and use the present partially developed picture for purposes outside science, do so at their own risk.

The lack of finality of scientific theories would be a very serious limitation of our argument, if we had staked much on their permanence. The religious reader may well be content that I have not offered him a God revealed by the quantum theory and, therefore, liable to be swept away in the next scientific revolution. It is not so much the particular form that scientific theories have now taken—the conclusions which we believe we have proved—as the movement of thought behind them that concerns the philosopher. Our eyes once opened, we may pass on to a yet newer outlook on the world, but we can never go back to the old outlook.

MYSTICAL RELIGION

We have seen that the cyclic scheme of physics presupposes a background outside the scope of its investigations. In this background we must find, first, our own personality, and then perhaps a greater personality. The idea of a universal Mind or Logos would be, I think, a fairly plausible inference from the present state of scientific theory; at least it is in harmony with it. But if so, all that our inquiry justifies us in asserting is a purely colourless pantheism. Science cannot tell whether the world-spirit is good or evil, and its halting argument for the existence of a God might equally well be turned into an argument for the existence of a Devil.

I think that that is an example of the limitation of physical schemes that has troubled us before—namely, that in all such schemes opposites are represented by + and −. Past and future, cause and effect, are represented in this inadequate way. One of the greatest puzzles of science is to discover why protons and electrons are not simply the opposites of one another, although our whole conception of electric charge requires that positive and negative electricity should be related like + and −. The direction of time's arrow could only be determined by that incongruous mixture of theology and statistics known as the second law of thermodynamics; or, to be more explicit, the direction of the arrow could be determined by statistical rules, but its significance as a governing fact "making sense of the world" could only be deduced on teleological assumptions. If physics cannot determine which way up its own world ought to be regarded, there is not much hope of guidance from it as to ethical orientation. We trust to some inward sense of fitness when we orient the physical world with the future on top, and, likewise, we must trust to some inner monitor when we orient the spiritual world with the good on top.

Granted that physical science has limited its scope so as to leave a background which we are at liberty to, or even invited to, fill with a reality of spiritual import, we have yet to face the most difficult criticism from science: "Here," says science, "I have left a domain in which I shall not interfere. I grant that you have some kind of avenue to it through the self-knowledge of consciousness, so that it is not necessarily a domain of pure agnosticism. But how are you going to deal with this domain? Have you any system of inference from mystic experience comparable to the system by which science develops a knowledge of the outside world? I do not insist on your employing my method, which I acknowledge is inapplicable, but you ought to have some defensible method. The alleged basis of experience may possibly be valid, but have I any reason to regard the religious interpretation currently given to it as anything more than muddle-headed romancing?"

The question is almost beyond my scope. I can only acknowledge its pertinency. Although I have chosen the lightest task by considering only mystical religion—and I have no impulse to defend any other—I am not competent to give an answer which shall be anything like complete. It is obvious that the insight of consciousness, although the only avenue to what I have called *intimate* knowledge of the reality beyond the symbols of science, is not to be trusted implicitly without control. In history, religious mysticism has often been associated with extravagances that cannot be approved. I suppose too that oversensitiveness to aesthetic influences may be a sign of a neurotic temperament unhealthy to the individual. We must allow something for the pathological condition of the brain in what appear to be moments of exalted insight. One begins to fear that after all our faults have been detected and removed there will not be any "us" left. But in the study of the physical world we have ultimately to rely on our sense-organs, although they are capable of betraying us by gross illusions; similarly, the avenue of consciousness into the spiritual world may be beset with pitfalls, but that does not necessarily imply that no advance is possible.

As scientists, we realise that colour is merely a question of the wavelengths of aethereal vibrations, but that does not seem to have dispelled the feeling that eyes which reflect light near wavelength 4800 are a subject for rhapsody whilst those which reflect wavelength 5300 are left unsung. We have not yet reached the practice of the Laputans, who, "if they would, for example, praise the beauty of a woman, or any other animal, they describe it by rhombs, circles, parallelograms, ellipses, and other geometrical terms." The materialist who is convinced that all phenomena arise from electrons and quanta and the like controlled by mathematical formulae, must presumably hold the belief that his wife is a rather elaborate differential equation, but he is probably tactful enough not to obtrude this opinion in domestic life. If this kind of scientific dissection is felt to be inadequate and irrelevant in ordinary personal relationships, it is surely out of place in the most personal relationship of all—that of the human soul to a divine spirit.

I am standing on the threshold about to enter a room. It is a complicated business. In the first place, I must shove against an atmosphere pressing with a force of fourteen pounds on every square inch of my body. I must make sure of landing on a plank travelling at twenty miles a second round the sun—a fraction of a second too early or too late, the plank would be miles away. I must do this whilst hanging from a round planet head outward into space, and with a wind of aether blowing at no one knows how many miles a second through every interstice of my body. The plank has no solidity of substance. To step on it is like stepping on a swarm of flies. Shall I not slip through? No, if I make the venture one of the flies hits me and gives a boost up again; I fall again and am knocked upwards by another fly; and so on. I may hope that the net result will be that I remain about steady, but if, unfortunately, I should slip through the floor or be boosted too violently up to the ceiling, the occurrence would be, not a violation of the laws of Nature, but a rare coincidence. These are some of the minor difficulties. I ought really to look at the problem four-dimensionally as concerning the intersection of my world-line with that of the plank. Then again, it is necessary to determine in which direction the entropy of the world is increasing in order to make sure that my passage over the threshold is an entrance, not an exit.

Verily, it is easier for a camel to pass through the eye of a needle than for a scientific man to pass through a door. And whether the door be barn door or church door it might be wiser that he should consent to be an ordinary man and walk in rather than wait till all the difficulties involved in a really scientific ingress are resolved.

Acknowledgments

Physics and Beyond by Werner Heisenberg. George Allen & Unwin, London.

Where Is Science Going? by Max Planck. George Allen & Unwin, London.

Ideas and Opinions by Albert Einstein. Copyright © 1954, 1982 by Crown Publishers, Inc. Used by permission of Crown Publishers, Inc.

"Positivism, Metaphysics and Religion (1952)" from *Physics and Beyond: Encounters and Conversations* by Werner Heisenberg. Translated from the German by Arnold Pomerans. Copyright © 1971 by Harper & Row, Publishers, Inc. Reprinted by permission of the publisher.

Mind and Matter by Erwin Schrödinger. Reprinted by permission of the publisher, Cambridge University Press.

My View of the World by Erwin Schrödinger. Reprinted by permission of the publisher, Cambridge University Press.

Science and Humanism by Erwin Schrödinger. Reprinted by permission of the publisher, Cambridge University Press.

Nature and the Greeks by Erwin Schrödinger. Reprinted by permission of the publisher, Cambridge University Press.

What Is Life? by Erwin Schrödinger. Reprinted by permission of the publisher, Cambridge University Press.

Science and the Unseen World by Arthur Stanley Eddington. Reprinted by permission of the publisher, Cambridge University Press.

New Pathways in Science by Sir Arthur Eddington. Reprinted by permission of the publisher, Cambridge University Press.

The Nature of the Physical World by A. S. Eddington. Reprinted by permission of the publisher, Cambridge University Press.

From *Physics and Microphysics* by Louis de Broglie, translated by Martin Davidson. Copyright © 1955 by Pantheon Books, Inc. Reprinted by permission of Pantheon Books, a Division of Random House, Inc.

"Scientific and Religious Truths" © Werner Heisenberg, *Schritte Über Grenzen*, R. Piper & Co. Verlag 1971; English edition *Across the Frontiers* published by Harper & Row, 1974.